T3-BRK-863

Making a Green Machine

Studies in Modern Science, Technology, and the Environment
Edited by Mark A. Largent

The increasing importance of science and technology over the past 150 years—and with it the increasing social, political, and economic authority vested in scientists and engineers—established both scientific research and technological innovations as vital components of modern culture. Studies in Modern Science, Technology, and the Environment is a collection of books that focuses on humanistic and social science inquiries into the social and political implications of science and technology and their impacts on communities, environments, and cultural movements worldwide.

Mark R. Finlay, *Growing American Rubber: Strategic Plants and the Politics of National Security*

Jill A. Fisher, *Gender and the Science of Difference: Cultural Politics of Contemporary Science and Medicine*

Gordon Patterson, *The Mosquito Crusades: A History of the American Anti-Mosquito Movement from the Reed Commission to the First Earth Day*

Jeremy Vetter, *Knowing Global Environments: New Historical Perspectives on the Field Sciences*

Making a Green Machine

The Infrastructure of Beverage Container Recycling

FINN ARNE JØRGENSEN

RUTGERS UNIVERSITY PRESS
New Brunswick, New Jersey, and London

TD
794.5
.J67
2011

Library of Congress Cataloging-in-Publication Data

Jørgensen, Finn Arne, 1975–
Making a green machine : the infrastructure of beverage container recycling / Finn Arne Jørgensen.
p. cm. — (Studies in modern science, technology, and the environment)
Includes bibliographical references and index.
ISBN 978-0-8135-5054-1 (hardcover : alk. paper)
1. Beverage containers—Recycling—Scandinavia. 2. Beverage containers—Recycling—United States. I. Title.
TD794.5.J67 2011
363.72′88—dc22

2010045434

A British Cataloging-in-Publication record for this book is available from the British Library.

Copyright © 2011 by Finn Arne Jørgensen

All rights reserved

No part of this book may be reproduced or utilized in any form or by any means, electronic or mechanical, or by any information storage and retrieval system, without written permission from the publisher. Please contact Rutgers University Press, 100 Joyce Kilmer Avenue, Piscataway, NJ 08854–8099. The only exception to this prohibition is "fair use" as defined by U.S. copyright law.

Visit our Web site: http://rutgerspress.rutgers.edu

Manufactured in the United States of America

For Marion and Lina

University Libraries
Carnegie Mellon University
Pittsburgh, PA 15213-3890

Contents

Figures

Preface

In Norway, recycling bottles and cans is an activity we more or less take for granted, and the machines we use for returning them tend to blend into the general shopping experience—except when they malfunction. This pervasive anonymity is what originally inspired me to study the history of reverse vending machines (RVMs). In attempting to discover how we got to this point, I found that RVMs were far from simple technologies placed in grocery stores. They are the front ends of large technical systems that harness billions of consumer environmentalist actions. At the same time, the RVMs and the systems are key components in fierce conflicts over beverage market shares. This intersection between business interests, everyday environmentalism, and technological development gives us a vital insight into the making of modern environmental policies.

This study focuses primarily on the history of the Norwegian company Tomra Systems, established in 1972, which produces most of the reverse vending machines on the world market today. While older machines for bottle returns existed, Tomra was the first company to use high-tech solutions, such as optical recognition, microprocessors, and laser technology, to solve the problem of beverage container returns. Although Tomra's founders developed their first machine for the Norwegian market, Tomra is now the world's leading producer of these machines, with 80–90 percent of the world market. Because of this dominant market position, a study focused on Tomra can reveal broad international trends in the development of systems for recycling our bottles and cans.

I have based this history on a wide range of sources, including Norwegian and international newspaper and trade journal articles; corporate annual reports and advertising material; a few internal documents; and a substantial amount of oral interviews with the Planke brothers, as well as former Tomra CEOs, scientists, and board members. I have also interviewed grocers who installed the Tomra machine in the 1970s and have consulted the literature from the grocers' trade associations in Norway. I have studied documents on early bottle deposit discussions and the Resirk recycling system in the Norwegian Ministry of the Environment archives and the National Archival Services of Norway and interviewed central actors from the consortium's history. Although I did not do physical archival research in Sweden or the United States—which are two critical locations in understanding the movement of the RVM—I was able to find a wealth of material (both governmental documents and media coverage) online and through libraries. Furthermore, I collected patent applications from the Norwegian University of Science and Technology (NTNU) library collections as well as through Google Patents. All translations from Norwegian or Swedish sources are my own.

How do the available sources and evidence limit and structure the narrative choices I have made? When one studies the marginal, the invisible, and the seemingly trivial, the evidence often takes on the same character. As a result, the evidence can be missing or disregarded. Tomra's reverse vending machine has many different users, and not all of them are consumers. Because of the source material available to me during the writing, much of my story is seen from the producer side. My decision to focus on Tomra's development of the reverse vending machine shaped my use of the material available to me.

Getting access to business archives and business records can be challenging for historians. Most companies, in particular those that have a share value to protect, manage their information flow very carefully. Even in cases where the companies are cooperative and give the historian more extensive access, the material that would be most fruitful for historical research may no longer exist. Indeed, it seems like most American business histories are written about a handful of large companies and entrepreneurs with large and well-ordered archives run by professional historians and archivists. Small and rapidly growing companies led by entrepreneurs often have other and more pressing concerns than preserving their documents for posterity. Although Tomra kindly gave me access to parts of their archives, most of the material I found there was published matter such as annual reports, marketing brochures, and newspaper clippings. I did not have access to board meeting minutes, product development reports, and similar internal documents.

I relied on the help and cooperation of many people when collecting the source material for this book. Petter and Tore Planke met with me on many occasions to discuss Tomra's history and provided me with stacks of historical material that I was not able to find elsewhere. Thanks to all my other interviewees—Aage Fremstad, Svein Jacobsen, Jørgen Randers, Stein Stugu, and Fredrik Tveitan—for their time. At Tomra I want to thank Bernt Saugen, who was my entrance point to Tomra (thanks to Stig Larssæther for putting me in touch with Saugen); Carol Quinn, who was my contact at Tomra when I started working on the project; and Andrew Young, my current contact.

Many people supported and inspired me during my work on this book. It started out as a PhD dissertation in the Department of Interdisciplinary Studies of Culture at NTNU. The key funding for my PhD fellowship came from the Productivity 2005 research program, co-financed by NTNU and the Research Council of Norway, and from the Faculty of Arts, NTNU. My advisors Per Østby and Stig Kvaal deserve thanks for suffering such a stubborn student for years. I have also benefited much from the feedback and criticism I got from my doctoral defense opponents Arne Kaijser and Thomas Zeller.

I spent a year of writing at the Department of Science, Technology, and Society at the beautiful University of Virginia. Thanks to the department chair, professor Deborah Johnson, for providing me with office space (as well as library privileges at the wonderful UVA libraries) during the 2005–2006 school year. During this year I worked with the Committee for the History of Environment and Technology and took some truly inspiring classes with Professors Jack Brown, Bernie Carlson, and Ed Russell—a special thanks to all of you!

Looking back on the years I've been working with this book, I realized that I have presented in-progress work from this book project at ten international conferences and workshops. Countless people deserve thanks for the comments and ideas they have given me, though I want to highlight Reggie Blasczyk, Thomas Brandt, Kjetil Fallan, Victoria de Grazia, Lise Kvande, and Chris Rosen.

Doreen Valentine was my first editor at Rutgers University Press, and I'm indebted to her for taking such a keen interest in my work. Peter Mickulas took over the responsibility for my book after Doreen accepted a new job offer and I'm grateful for the work he put into bringing the book to print.

Finally, I need to thank my family for all their help and support. While it is not always easy to combine family life and its complexities with an academic career, one thing is for sure: I could not have done it without you. This book owes a tremendous amount to my wife, Dolly, whose sharp attention to detail and grasp of the big picture has been invaluable. Thank you for inspiring me, rooting for me, and taking care of me!

The Companion Website rvmhistory.org

This book comes with a companion website, where it will be possible to read an illustrated essay on the visual history of RVMs, watch the videos and advertising material referenced in the book, and find links to key texts in the historical debate on recycling. The site will complement the written text by providing full-color illustrations and other multimedia material that is hard to publish in print, as well as provide material that will make the book suitable for classroom use. Furthermore, visitors will be able to contribute personal stories about their own experiences with particular RVM models. The website can be found at http://www.rvmhistory.org/

Making a Green Machine

Chapter 1 Bottles, Cans, and Everyday Environmentalism

A HUNDRED YEARS AGO, A BEER BOTTLE had much in common with a library book. Like the book, a bottle did not belong to the consumer, but was rather a valuable resource that belonged to the distributor. A deposit paid upon purchase encouraged consumers to return the bottle to the store where they had bought it; in some cases, in fact, you could not buy more beer without returning bottles at the same time. When the consumer returned the bottle to the store, the grocer would refund the deposit. The bottles would then be sent back to the bottler, typically upon delivery of full bottles to the store. A single bottle could thus be refilled dozens of times, becoming part of a closed loop between bottler, distributor, and consumer. Today, the situation has become much more complicated. The closed loop can no longer be taken for granted. We have learned to think about beverage containers in particular ways in today's world of digital downloads, global beverage brands, and disposable packaging. We now tend to think about our own experience of the product and the content, and are less concerned with the material delivery mechanisms. Beverage containers are no longer valuable resources. Yet the sheer mass of hundreds of billions of beverage containers warrants a closer look at the history of what we do and have done with them.

The history of beverage containers perfectly encapsulates the development of modern consumer society. Increasingly affluent consumers purchased ever-larger quantities of bottled and canned beverages in the post–World War II years. Policy historian and activist Frank Ackerman has argued that the idea of refillable glass bottles "had to be unlearned in order to unleash the consumer society of the late twentieth century."[1] As the beverage industry grew from predominantly local companies into global corporations selling hundreds of billions of beverage containers every year, something had to give. In most cases, this meant the end of refillable bottles and deposit systems. However, this historical process of substituting refillable bottles with disposable containers of various types and materials did not happen without resistance. Nor can we take the outcome for granted.

Worldwide, we can find a wide variety of solutions to the bottle problem. Let me give you an idea of some of those systems through four glimpses into today's bottle recycling cultures.

CHARLOTTESVILLE, VIRGINIA, OCTOBER 2005. On a walk, my wife and I and our dog walked across a small stream on one of the many small trails surrounding our town. Under the bridge, swirling in green foam, several plastic water bottles bobbed up and down in the water. Virginia is just one of the thirty-seven American states that do not have a law requiring some kind of deposit system for recycling bottles. As a good environmentalist, I would gather our own bottles (mostly glass microbrews, plastic water bottles, and my wife's Coke aluminum cans) as well as high-density polyethylene (HDPE) juice containers and drive over to the McIntire Road Recycling Center to return them. The recycling center accepted most forms of containers, but occasionally refused to receive certain types, lacking demand for the recycled material. Still, returning my containers here made me feel slightly better about my own consumption habits.

BEIJING, CHINA, JULY 2008. In an effort to make the Summer Olympic Games greener, Chinese authorities moved more than one hundred thousand of Beijing's garbage pickers out of town. China does not have any formal beverage container recycling system. Instead, garbage pickers have collected bottles and cans—along with other waste—from the street to sell them to recycling stations. While these workers contribute to one of the most effective recycling economies in the world, they did not fit into Beijing's campaign to present a modern, sanitized city to the world during the Olympics.[2] Nor do they fit Western conceptions of greenness and recycling. The hordes of Western Olympic visitors in Beijing preferred both their trash and their recycling systems to be invisible—out of sight, out of mind.

NEW YORK, APRIL 2009. Environmentalists cheer as the Bigger Better Bottle Bill finally passed on April 3, 2009. New York is one of eleven U.S. states that currently has an active bottle bill, along with California, Connecticut, Delaware, Hawaii, Iowa, Maine, Massachusetts, Michigan, Oregon, and Vermont. Almost thirty years after the creation of the state's first—and flawed—bottle bill, nine years of intense lobbying from both opponents and proponents seem to have come to a conclusion.[3] Under the old bottle bill, only beer and soda bottles and cans were required to carry deposits. Since the creation of the law, however, new bottled beverage types have completely taken over the market. The main result of the new law is to also cover the now ubiquitous single-serving plastic bottled-water containers. While the New York Bigger Better Bottle Bill shows that changes are possible, there is definitely strong resistance to bottle bills, despite their efficiency in increasing redemption rates. The eleven deposit states have an average redemption rate of 61.4 percent for all containers, while the average rate for nondeposit states is only 24.2 percent.[4]

TRONDHEIM, NORWAY, DECEMBER 2009. Recycling empty bottles and cans is an integral part of daily life. In a country of fewer than five million inhabitants, more than a million deposit bottles and cans are returned to grocery stores every day.[5] Hundreds of millions of kroner are paid in deposits and later refunded to the consumers. This everyday recycling is a seemingly invisible and insignificant activity, yet the scale and scope of this system extends wide and deep into Norwegian society. Homeless people dig through thrash cans to find discarded bottles. Kids collect bottles to buy candy. Young people bring back empty beer cans after last weekend's parties. Families bring full plastic sacks of bottles when doing their Saturday grocery shopping. Even a billionaire like the finance guru Trygve Hegnar recycles his bottles, not because of environmental issues, "but because it is about money."[6] But it is also about convenience and technology; container returns take place at the point of purchase, so consumers do not need to make a trip to a separate recycling station. Consumers return almost all the beverage containers through reverse vending machines (RVMs) in grocery stores, getting a refund receipt in return. The RVMs have an unassuming look; they are basically a hole in the wall with a small display showing the number of bottles returned, the total refund amount, and any error messages should they arise. As the bottles and cans pass through this hole in the wall, they become part of a large, technological system set up to recycle and reuse these containers. The system is highly successful; in 2008, 99 percent of refillable bottles, 91 percent of aluminum cans, and 89 percent of plastic PET bottles were returned.[7]

These four vignettes from the world of beverage container recycling tell us several things about the relationship between technology, business, and the environment.

1. Small things can have big effects. Bottles seem trivial and unimportant, but they are far from it. Consider an empty bottle, one of the hundreds of billions of beverage containers emptied and discarded throughout the world every year. It is a remarkably complex object, materially, culturally, and politically. The containers our beverages come in are not just physical objects made of glass, aluminum, steel, or plastic. As we will see in this book, they are also cultural objects—signifying and projecting meaning about their contents, their producers, and their consumers. They are political artifacts, generating heated controversies between environmentalists, industry, and policy makers. They are highly engineered technical artifacts, some as an evolutionary result of thousands of years of history, while others were created in fiercely competitive and intense R&D projects. When emptied, freed of their original function, they become a problem—or a resource. To paraphrase one of Melvin Kranzberg's laws of technology, bottles are neither good nor evil, but nor are they neutral.[8] They are at the center of large business decisions, influencing the setup of huge transnational business systems. A billion individual bottles are just trash. But put them in a system, and they become resources. This, however, requires a culture of recycling, technology, and business cooperation.

2. *Because the aggregated financial and environmental stakes of these small things are so high, the discussion is often polarized.* Business groups often come out as the bad guys in beverage container recycling controversies. However, "business" is not a large, monolithic entity with common interests. First, the relationship between large and small bottlers, local and national ones, has been little explored. Even among large bottlers, interests and approaches vary, as Coors's involvement in the development of aluminum can recycling in the 1950s and the California bottle bill in the 1980s demonstrates.[9] Second, beverage companies are far from the only business stakeholders in this story. The aluminum industry, beverage container producers, and large and small grocers all enter the beverage container story at different stages. Labor organizations and interests have taken a particularly dualistic stance to changes in beverage container technologies. As a result, recycling and waste management questions have become highly controversial and reach far beyond simple business interests. The key issues are, first, whether recycling should be mandatory or voluntary and, second, whether consumers or producers should be responsible for waste and recycling. Free-market evangelists argue that recycling should occur only when entrepreneurs find it worthwhile and profitable to recycle certain materials.[10] Others argue that society has a responsibility for encouraging—or enforcing—recycling for environmental, conservation, or aesthetic reasons.[11] The interesting point to me is not whether bottle and can recycling falls down on one side or the other of this largely ideological divide, but rather how ideas of recycling and resource management have constantly been reinterpreted by different groups, at different times, in different locales, and in light of new forms of environmentalism.

3. *While the same beverage companies are active all over the world, they meet different jurisdictions in different areas*. As a result, a bottle has come to represent highly different things in different locations. This book will focus closely on how recycling technologies and ideas travel across the borders—geographic, material, economic, and cultural—between business, technological and political systems, and on the forces that limit the unrestrained use of disposable containers. The book follows beverage containers, corporations, and recycling technologies as they move between markets. In some places the beverage companies and their allies have gained tremendous influence on national political systems, often to protests from nongovernmental organizations (NGOs).[12] In other places, national environmental legislation has curtailed the actions of multinational beverage companies. We will see examples of how environmental legislation has served as protectionism. We will also see how globalization allows technological and cultural solutions to migrate across borders—but not unchanged and not without resistance. Most of the RVMs that appear in these different systems come from a Norwegian company. The RVM developed in a particular national recycling system, but yet not in a complete vacuum. Empty beverage containers were transformed from a local littering problem to a global environmental problem from the 1960s to the 1990s.

4. *Technological systems have become a key part of the solutions*. In the Western world, bottle recycling has moved away from the manual garbage pickers that dominate China—and part of our history, too—to extensive technological infrastructures that make recycling convenient and economical for consumers. The RVM stands at the center of so many of these systems, in particular the ones with high recycling rates. Behind the hole in the wall is a large technological system orchestrating environmentalism. Just as Thomas Hughes has illuminated the multiplicity of actors and interests involved in building electrical systems into the obdurate systems penetrating modern society, the beverage container recycling systems have developed through complex and sometimes controversial negotiations between technological possibilities and political goals.[13]

THESE FOUR LESSONS DEMONSTRATE that empty bottles are vessels into which many discourses can fit. Throughout this book, we will see how they are filled, emptied, and refilled with meaning, values, and politics. I aim to go beyond a story of just environmental politics, of organizations and businesses making and influencing legal frameworks. The story I tell firmly embeds technology in the making of modern environmentalism. While beverage container recycling is a commonplace and unremarkable everyday event, the systems that make recycling possible were created where business, politics, and technology met.

When emptied, beverage containers no longer fill their role of containing liquids and potentially become trash. This book is primarily concerned with historical attempts to develop technological and organizational systems that can keep bottles from becoming waste. Previous histories of recycling and waste management have to a large extent attempted to expose the underbelly of garbage disposal—what happens to your trash after you throw it away—rather than attempt to look at the prevention of waste.[14] A second type of trash scholarship has focused on the historical development of waste disposal services: what were the means and motivations for handling garbage.[15] Third, scholars have looked at how cultural conceptions of waste have changed over time: what is waste to us now were valuable resources to previous generations—and vice versa. In this interpretation, garbage can tell us something about the reigning norms and values of a culture.[16] This last perspective is certainly relevant for the history of beverage container recycling, since new consumer attitudes, production methods, materials, and distribution systems have changed the rules of the game completely in the past one hundred years.

Today we often take it for granted that the goal of recycling and waste prevention is to preserve the environment. It has become a way consumers can connect their everyday activities with the global environment. Today's environmentalism runs shallow and wide. Instead of addressing environmental concerns by radically changing their lifestyles and reducing their consumption, most people engage in small acts of *everyday environmentalism*—like recycling bottles, paper, plastic, and glass; using low-energy lightbulbs; occasionally biking to work; carpooling if they drive (preferably in a hybrid car); buying ecological produce every once in a while; buying carbon credits to offset the environmental cost of the seasonal vacation flight; or being a passive member of an environmentalist organization—while at the same time maintaining or increasing their consumption level.

In his study of the rise of French environmentalism and its impact on French society, Michael Bess argues that the outcome of the radical environmentalism of the 1960s and 1970s is a light-green society, a blend of environmentalism and technological enthusiasm that resulted in a "partial greening of the mainstream, in which neither side emerged wholly satisfied, not utterly dismayed, but in which a whole new complex of discourses and institutions nonetheless came into being."[17] Instead of leading to a radical green society, environmentalism seems to have become just another choice in the postmodern supermarket of modern consumerism. One might say that the light-green society is what we got when the radical environmentalists of the counterculture generation cut their hair and got a job.

Bess's argument can be read as either a success story, in which green ideas gradually came to permeate mainstream culture and economy, or as a defeat story, wherein virtually all the radical aspects of the original green vision were trimmed down and ignored by a consumerist population. Many environmental historians have chosen the latter interpretation, electing to see modern environmentalists as hypocrites. In the words of Deborah Lynn Guber, "Environmental concern is widespread, behavior is not."[18] Hal Rothman has similarly portrayed Americans as halfhearted environmentalists who are green "when it is inexpensive—economically, socially, and culturally," but who at the same time are reluctant to "collectively sacrifice convenience and even the smallest of material advantages to assure a 'cleaner' future."[19] Such interpretations are clearly of a declensionist character, asserting that there is a large gap between environmental ideals and consumer lifestyles. Al Gore argued for the futility of isolated, personal actions, such as changing a lightbulb or returning bottles; in the light of climate change, only massive, structural efforts are effective.[20]

Instead of taking such a fatalistic approach, this book examines the development and effectiveness of the system recreated to enable one of these isolated environmental actions—recycling bottles and cans. The everyday environmentalism that resulted from the mainstreaming of the radical environmentalism of the 1960s and 1970s does not stem solely from grassroots idealism, but rather was born at the heart of capitalist production and distribution. As I have indicated, this is a kind of environmentalism that does not require much effort or sacrifice from the individual consumer; convenience is one of its more prominent qualities. Everyday environmentalism is in league with, and not opposition to, consumer society. Likewise, it exists in a symbiotic relationship to technology.

Scholars have largely ignored the large sociotechnical infrastructures that work toward making everyday environmentalism like the recycling of bottles and cans effective as an environmental measure instead of just a collective green illusion. In so doing, they have failed to recognize the role of business and technology in promoting environmental goals. We need to understand why businesses engage—or do not engage—in recycling, and how their activities intersect with consumer recycling. Like consumers, beverage producers want convenience as well as to reduce risk and insecurity. They are not necessarily evil, though environmentalist descriptions can occasionally make it seem so. It is true that the consumer's involvement with beverage container recycling usually stops in the grocery store, where the empty bottle or can is fed through a hole in the wall. But unless we consider the larger sociotechnical system that includes policies, businesses, technologies, and cultural constructs behind the hole in the wall, we miss out on most of the story. We only see the tip of the landfill, so to speak.

Viewing technology in a system perspective allows us to look beyond the machine toward the society it is integrated in. Hughes has argued throughout his scholarship that technologies must be analyzed in the context of the political, economic, and social factors that influence them; they are embedded in large technological systems.[21] Technology does not develop isolated from society, but rather in parallel, as part of a seamless web. For instance, the car cannot be meaningfully analyzed as an isolated artifact, but rather must be seen as part of a larger system consisting of manufacturers, service workshops, roads, credit institutions, and so on. However, the system itself is more or less invisible; we only see the individual artifacts that constitute the system. We should thus consider the system as a model that can connect relevant factors to each other.

In order to comprehend how these systems come into being, Hughes advocates studying the system builders. These are the entrepreneurs who attempt to influence their surroundings so that the system can reach optimal performance. In Hughes's narratives, his historical actors had to handle a series of problems, called *reverse salients*, when building their systems. Many of these problems were nontechnical in nature: questions about economy, politics, or organizations could be equally important to the system's development. As Hughes tells the story, many systems often started out as an invention. But in order for an invention to turn into a successful product, it must be developed from something that "works" in the inventor's laboratory to something that functions in society. During this process, the system builders often move the technology through several phases of increasing complexity before the technology gains sufficient *momentum* to influence its surroundings. This momentum can be seen as a way of describing the interplay of the obduracy of the built environment, vested interests, and culturally accepted solutions to certain problems. For instance, the automobile system resists change because of the many interests groups that benefit from the current system; it encourages continued road construction and fossil fuel consumption. Technologies like the electric car challenge this system, which is one reason why these have failed to become successful on a large scale.[22] Through his systems model, Hughes argues that technologies never work in isolation. They are in fact composed of multiple entities: the inventor/innovator, businesses, users, and the environment, all operating within a cultural framework.

This takes us into a complex territory shaped by organizations that aim for environmental goals far removed from modern consumerism, by consumers who want to do their bit for the environment while at the same time living a good life, companies that recognize environmental concerns but do not know whether they will lead to corporate profits, and finally governmental policy makers who have to balance all these concerns.

THIS BOOK EXAMINES THE MODERN HISTORY of empty beverage containers, how they have continually been redefined as economic and environmental problems, and how we have attempted to control them. My goal in this book is to trace the development of beverage container recycling systems and the technological infrastructures developed by business to facilitate and support these systems. I examine the parallel technical development of reverse vending machines (RVMs, used for consumer returns of empty bottles and cans) and the cultural context of beverage container recycling. I analyze how industry, policy makers, environmentalists, and consumers have handled and discussed empty beverage containers from the 1960s through the 1990s in Norway, Sweden, and the United States. I have selected these areas because the United States led the way in the adoption of disposable containers and Scandinavia pioneered the development of technical and political infrastructures to handle the containers. The juxtaposition of the two areas allows us to see how the problems developed and how potential solutions might be implemented.

This book thus examines the interlinked stories of beverage container recycling infrastructures in the United States and Scandinavia to see how these systems developed. Our main focus will be on Norway and the Norwegian company Tomra, which is now the world's leading producer of RVMs for beverage containers. This company played a vital role in the development of viable technical solutions in both Norway and Sweden (and tried to do the same in the United States). The RVM is a key technology in the Scandinavian recycling system. Placed in most grocery stores, it allows consumers to return empty bottles and cans through a hole in the wall, and gives back a deposit-refund slip. Because of their central position, we will focus closely on Tomra's foundation, marketplace struggles, and eventual marketing of their RVM as a green environmental technology.

In Chapter 2 we will look at how businesses interact with political systems. It begins with an overview of the manual system of recycling in place when reusable glass bottles dominated the marketplace. In this system, all bottles carried a deposit that was refunded when the bottle was returned. This system was eventually replaced by disposable containers without deposits in the United States in the 1950s and 1960s. The same trends occurred in Scandinavia, but in contrast to the American story, market forces were not allowed to freely discard the previously established recycling practices.

Chapter 3 moves into a story of technological innovation with a discussion of how the RVM was designed to meet specific infrastructure needs of Norwegian grocers, consumers, and bottlers. As the machine moved into the global marketplace, its designers had to respond to the challenges of new and varied container types. Chapter 4 looks at the redesign of Tomra's RVM to handle multiple glass bottle types on an international market, whereas Chapter 5 moves into the realm of aluminum cans. Through a comparison of what happened with can recycling in Sweden and New York, we get a better understanding of how recycling culture both influences the success of the system and is part of the system.

Whereas beverage container recycling started out as an infrastructure issue, by the late 1980s it had been linked to environmentalist concerns. Chapter 6 looks at the greening of the RVM: how the machine and recycling systems helped consumers become good environmental citizens. This discussion continues in Chapter 7, which examines the political controversy over allowing disposable containers without tariffs in Norway and how this activity had to be linked with environmental goals. Since Norway already had an extremely successful bottle recycling system in place, the only way disposables could be introduced was as part of a full-scale, environmentally friendly recycling system.

Chapter 2 The Problem of Bottles

IN THE FALL OF 1961 THE NORWEGIAN ACTOR and comedian Carsten Byhring ridiculed the new one-way, nonreturnable beer bottle at the Edderkoppen Theatre in Oslo. As he put it, the problem was not that the distinctively chubby bottle looked so bad, at least any more than other empty bottles. But what were you supposed to do with it? You were not allowed to leave it anywhere. You could not sell it, or even give it away for free. No one knew how to get rid of it—it was yours forever. "Carefully wrap the bottle in paper and throw it down the garbage chute," said the packaging. "But who the hell carries a garbage chute around," quipped Byhring. In the good old days, having an empty bottle gave you a thirty-øre refund—and then you were well on your way to a new bottle of beer.[1] "You kind of were part of the traffic flow then, you belonged in the trade, in the run between the brewery, the store, and yourself, going one way, and then yourself, the store, and the brewery going the other way—a kind of interdependence. And now you're all alone. Back then, you had a kind of a sense of responsibility, something to take care of in the community—the empty bottle."[2]

Byhring's performance is wonderfully illustrative of the place glass bottles had in Norwegian society in the 1960s. They were far more than simple beverage containers. First, bottles were part of an organizational system set up to handle distribution and returns. Bottlers and brewers bought back used bottles from consumers through a deposit-refund system. This system was maintained in a largely cooperative relationship between these groups.

Second, Byhring reminds us that this was a time when the beverage industry began looking for new and, for them, more rational ways to distribute their products. The geography of Norway presented the brewery industry with unique distribution challenges. Norway is a very long and skinny country with a dispersed population. Expanding beyond local distribution chains was expensive, primarily because the use of returnable bottles required the bottles to be transported back to the brewery. The one-way container held a promise to solve these problems. It required neither transport back to the brewery nor cleaning and refilling. For this reason, some of Norway's largest breweries made plans to invest in expensive canning facilities. At the same time, when Norway joined the European Free Trade Area (EFTA) in 1960, international competition increased. Foreign breweries could now send their beer to Norway duty free. By maintaining the bottle deposit system, Norwegian breweries could prevent some of this competition, because foreign bottles would be excluded from the scheme. Bottlers and brewers were thus torn in two directions with regard to deposits and disposable bottles.

Third, bottles were part of a social system specifying how empty bottles were to be treated—a norm ingrained in all Norwegians, even the beer-drinking bum that Byhring imitated. Postwar Norway can be seen as a country infused with a Puritan ethic of saving money and working hard.[3] People cared for the empty bottle because of the deposit—after all, you can't throw away money! When the deposit disappeared, they were at a loss about what to do with the bottle, not unlike my own reaction when moving to Virginia in 2005.

Fourth, Byhring's laments also hint that this social structure—and thus Norwegian society—was changing. As Norway moved toward an increasingly affluent consumer society, the old networks and structures that regulated both social behavior and physical matter began dissolving under considerable pressure. Consumerism and hedonism threatened the traditional, sober Norwegian mentality. After the government permitted free ownership of cars in 1960, a new leisure-oriented lifestyle spread.[4] During the 1960s, car ownership quadrupled from 220,000 to 846,000. Convenience and individual freedom converged in the car, making Norwegians increasingly mobile. Increased marketing and the new mobile lifestyle helped to make bottled beverages more popular. One consequence was that the consumption of beverages now took place in new settings: along roads and in nature. In the outdoors it could be hard to find a proper place to dispose of empty bottles, as Byhring reminded us. Bottles were originally an economic and logistical problem, but they turned into a littering problem during this period.

In this chapter we are going to look at how the material constraints and opportunities of using glass bottles shaped the postwar beverage industry. As disposable beverage containers became available in the 1960s, we will see how the existing Norwegian glass container recycling systems came under pressure. The bourgeoning environmental movement of the early 1970s led to a linkage between disposable containers and littering. This motivated policy makers to push against the threat of dismantling the existing glass recycling systems. Because of the nature of the issue, a full-blown controversy erupted in the Storting (the Norwegian parliament) over how to regulate disposable containers.

The Material Culture of Bottles

While empty bottles may at first seem a trivial and negligible problem, they have been the cause of ongoing controversies and negotiations between a series of actors and interest groups. The sheer physical materiality of hundreds of millions of empty bottles affected both nature and society. The problem of bottles reached deep into industrial, political, and everyday spheres. We need to understand how these issues developed. Only then can we comprehend how empty beverage containers have been transformed from a local littering problem to a global environmental problem—and consequently, how the technological infrastructures developed to handle the containers extended their reach far into everyday life, giving shape to new forms of both business and consumer environmentalism.

Bottled beverages really became ubiquitous in the postwar period. For example, in Norway, 825 million bottles were sold in 1973.[5] Reliable statistics for the American market are hard to come by, but in the early 1970s one study claimed that 5 percent of the solid waste in the United States consisted of containers produced by Coca-Cola.[6] The volume of bottles in the marketplace and their physical characteristics directly affected society and the environment. The environmental and cultural impact of this mass of materials cannot be underestimated. Neither can the challenges inherent in creating an infrastructure for the handling of these containers.

In the 1960s and 1970s, glass was still the dominant beverage container material in the world. Glass bottles are the oldest beverage containers still in use. No one is certain where, when, or how glass originated, but the estimates vary from 8000 BC to 2000 BC.[7] The ancient Romans perfected the glass bottle as a beverage container. Transparent wine bottles allowed their content to be seen and served to enhance the attractiveness of the beverage. Glass is made by combining sand, soda ash, and limestone; mixing it with crushed glass, called cullet; and firing the mixture in a furnace to a temperature of fifteen hundred degrees Celsius. For many beer bottles, carbon and sulfur is added to this mix to create the amber color that protects the beer from ultraviolet light, which makes it smell bad. The red-hot, molten glass is cut with shears into a gob, which then is made into a bottle. The bottles were handblown until early 1900s, when automatic systems began taking over this work—and thereby increasing production dramatically.[8]

The automatic mass production of bottles allowed beverages to be sold in packaged form to consumers. Earlier, beer and soda was mostly consumed at the point of purchase. When consumers could take the bottles home, distributors faced new problems. First, their distribution networks had to become even larger in order to reach the consumers. Second, the bottles were now spread over a large area. The challenge was how to get these bottles back for refilling. New bottles were expensive to produce. Bottlers and brewers attempted to solve this problem by offering to buy the bottles back from the consumers instead. According to the international trade magazine *Beverage World*, beverage producers organized deposit programs to achieve this goal by the early 1900s.[9] As far as I can tell, Norwegian bottlers and brewers started their deposit programs even earlier.[10] Most Norwegian brewery histories merely state that the deposit programs started "a long time ago." The refund system was desirable because it was cheaper for them to reuse bottles than to purchase new ones.

The way that glass bottles were used—and reused—put certain physical demands on the material. In order to handle the strain of transportation, handling, and cleaning, reusable glass bottles have to be fairly thick and solid. This also makes them very heavy. The high transportation costs served to limit distribution to local areas. Bottles as technological artifacts thus encouraged a decentralized bottling industry infrastructure, with many small breweries serving a local or regional market.[11] As we will see, when regional producers attempted to enter markets further afield, they began the search for alternatives to the heavy glass containers. This happened in both the United States and in Scandinavia, but here we will look in detail at what happened in Norway.

The Early Norwegian Deposit System

Historically, all Norwegian bottlers and brewers operated deposit systems to buy back their used bottles. While this was cheaper than making new bottles, they saw the work of handling the return bottles as inconvenient and ineffectual. It certainly placed large demands on the bottling facility infrastructure. For instance, Frydenlund Brewery in Oslo built a new empty-bottle storage building in 1936 that could store one million empty bottles, demonstrating the large volumes of bottles passing through the bottling facilities.[12] The Trondheim brewery E. C. Dahl's one-hundredth-anniversary history from 1956 graphically described the inconvenience of receiving used bottles, highlighting the large amounts of space necessary, as well as the difficulty of mechanizing the bottle handling.[13] In short, the logistics of bottle handling posed a large infrastructural challenge to breweries that increasingly strived for more efficient and cost-effective logistics.

Cleaning bottles was a labor-intensive job. The bottles had to be thoroughly washed in the bottling plant before they could be refilled. In 1899, the women—because this was a woman's job—who washed, filled, and capped bottles were paid thirty øre per one hundred bottles.[14] This meant that they had to clean four hundred to five hundred bottles in order to reach the average daily income for working women. Some bottles needed extra cleaning, since it was common to store kerosene in them; stores sold kerosene wholesale and empty bottles were a convenient way for customers to bring it home. Some of the women had as their job to smell all the incoming bottles, and if they caught as much as a little whiff of kerosene, the bottle had to be sent to special cleaning.[15] Machines later took over much of the cleaning work, but the inspection process still required people to ensure that the bottles were clean.[16]

These bottle-handling issues prompted bottlers to look for alternatives to reusable bottles. Yet in 1956 neither cans nor disposable bottles presented a viable alternative in Norway, primarily for economic reasons. Bottles were so expensive that not to return them would make the product excessively expensive for the consumer. However, the ingrained culture of reusing bottles was just as important as the economics of beverage containers. The previously mentioned Frydenlund anniversary book described American disposable bottles as bottles "that are sold together with the beer." The author's choice of words demonstrates the novelty of the idea of selling the bottle as well as the contents in 1950s Norway, where most people thought of the bottle as the property of the brewery. The writer described canned beer as something that had potential, but that required such large sales to be profitable that the conditions were "probably not very good in this country."[17] Until beverage container technologies improved or the culture of beverage consumption changed, Norwegian bottlers and brewers were stuck with reusable glass bottles.

While the brewers and bottlers competed over a small market, reusable bottles encouraged a remarkable cooperation. The different breweries maintained a common bottle stock of standardized bottles. Their common interest in reusing bottles made such a system rational. The Norwegian bottle types for thirty-five and seventy centiliters were standardized, based on discussions in the Norwegian Brewery Association about every ten years, starting in 1906.[18] The Brewery Association had a long history of concord and cohesion that was galvanized during resistance against a Nazi takeover of the association during World War II.[19] All the members of the association committed to using these bottle types, even though a few smaller breweries got an exemption. The size, color, and design of the bottles were continually adjusted to keep up with new technologies for capping and distribution. The standard in 1967 were the brown thirty-five- and seventy-centiliter bottles for beer and a twenty-centiliter soda bottle.[20] This allowed the breweries to reuse each other's bottles. In 1969, thirty-five-centiliter bottles were reused twenty-three times and seventy-centiliter bottles eighteen times, on average.[21]

For the consumer, much of the deposit system was invisible. When a consumer purchased a bottle, a small deposit was added to the price. The grocer refunded this deposit when the customer returned the bottle to the store. In some cases, you could not buy bottles without returning some used ones. Bottle returns were only a small inconvenience, and one that most people were accustomed to. The return rate was accordingly high. This reminds us how bottle returns were just as much a social system as an infrastructural system. Returning the empty bottles felt like a natural part of everyday life, as Byhring illustrated. This system was still in place in the early 1970s, even though changes in the production and distribution of beverages put it under considerable pressure beginning in the 1960s.

IN THE 1960S, TWO NEW BEVERAGE containers posed challenges to the existing deposit system: the disposable short "stubby" beer bottle and the reusable one-liter soda bottle. Disposable beverage containers held a promise for lower distribution costs for the largest breweries, leading them to try out such containers in Norway. Since the disposable bottles did not have to be reused, they could be made of lighter and thinner glass, something that decreased transportation costs. In 1960, the "stubby"—a short, disposable thirty-five-centiliter glass bottle for beer—appeared in Norwegian grocery stores. This was the bottle that Carsten Byhring ridiculed in 1961. The stubby became common roadside litter and most people considered it a nuisance. This triggered a discussion about the use of disposable containers in Norway.

Norway was not the first country to go down this path, but seemed set to follow in the footsteps of the United States. The United States was the origin of disposable beverage containers, global beverage brands, and consumer environmentalism, so we need to understand the Americans' attitudes to their empty bottles and cans.

Before World War II, soft drinks and beer in the United States were packaged mostly in refillable bottles, just as in Norway. The beverage industry maintained a deposit-refund system to buy back these bottles. Beverage container legislation requiring deposits would have been unnecessary at this time; the refillable-bottle system was an established industry practice and a fact of life for consumers.[22] In the 1960s, however, disposable containers became increasingly common. The market share of beer and soft drinks in refillable containers plummeted from over 90 percent in 1950 to 17 percent for beer and 34 percent for soft drinks in 1975, demonstrating that consumers embraced the new and more convenient containers.[23] Returning empty bottles to the grocery store gradually disappeared as an everyday consumer habit. Because of rising demand for throwaway containers, beverage production facilities changed in a way that made the move to disposable containers hard to reverse. Cleaning and reusing bottles demanded a completely different production infrastructure from that of disposable containers, so once the switch was made, it wasn't going to be changed. The move toward one-way containers also influenced the makeup of the industry. One-way containers benefit large producers and large distribution networks. Technological changes in packaging thus enabled large changes in the beverage industry, where dominance shifted over to the largest corporations.[24] The most visible result of this development was increased littering of empty beverage containers.

American industry groups took a stance in favor of recycling, but only for voluntary efforts. Mandatory recycling, taxes, or deposits implied producer responsibility for littering. In order to avoid such limits on their activities, industry groups initiated and sponsored environmental organizations like Keep America Beautiful (KAB) and the National Center for Resource Recovery. KAB was established as a nonprofit public-education organization in 1953 by a group of glass, aluminum, paper, and steel container manufacturers, including Coca-Cola, Pepsi, Seagram's, DuPont, Dow, and Procter and Gamble.[25] The National Center for Resource Recovery's board of directors was a "who's-who list" of American packaging companies.[26] Mirroring the National Rifle Association's credo, that of these organizations claimed that "packages don't litter; people do"—thus denying any responsibility for littering—and funded information campaigns to change consumer attitudes to littering and to raise environmental awareness.[27] An example of this is the 1963 film *Heritage of Splendor*, an eighteen-minute call for preserving the United States' natural resources from littering. The Ritchfield Oil Corporation produced the film for KAB and highlighted how American business demonstrated a great responsibility for nature through their wise use of natural resources. The actor Ronald Reagan's voice intoned how packaging became litter only after people thoughtlessly discarded it and that we needed collective and individual education to prevent littering.[28] KAB has to this day been consistent in arguing for information campaigns instead of legislation and taxes.

In spite of these industry efforts, several states passed bottle bills to counter the increased littering of one-way containers in the 1970s and 1980s, while others considered similar legislation. Oregon was the first state to do so in 1971, and then Vermont followed in 1972, Michigan and Maine in 1976,

Iowa and Connecticut in 1978, Delaware and New York in 1982, Massachusetts in 1983, and finally California in 1986.[29] Five-cent deposits became standard in the deposit states, other than Michigan, where nonrefillable bottles and cans carried a ten-cent deposit.

These bottle bills did not pass without resistance. The beverage industry was powerful, and it fought fiercely against bottle bills. They saw the bills as a "direct and politically motivated infringement on the free market and a threat to profits," according to Matthew Gandy.[30] William F. May of the American Can Company stated in 1976 that "we must use every tool available to combat bottle referendums this year in Maine, Massachusetts, Michigan and Colorado where Communists or people with Communist ideas are trying to get these states to go the way of Oregon."[31] The bottle bills were thus presented as a threat to free enterprise—to the very soul of the American way of life. According to Louis Blumberg and Robert Gottlieb, industry groups started an opposition campaign to smear the bottle bill by "every means possible."[32] These groups carried enough weight to stop or severely limit bottle bills in many states.

At the same time, it was difficult for industry to be openly against environmental protection. The industry-funded environmental organizations illustrate that industry actively worked to shape environmental policy and activity. Although the grassroots environmental organizations were involved in these discussions, business interests came to be much more influential in shaping recycling policies.[33]

American-style consumerism as well as its effects was hotly contested in postwar Norway. One of the recurring themes in the discussion about beverage container recycling was how to prevent Norway from ending up like Sweden or the United States, where littering of empty containers had become more common by the 1970s. In this respect, these countries worked as a vision of the future whereby Norwegians could see the consequences of not maintaining the bottle deposit system. It also indicates the ambivalent relationship Norwegians had to consumer culture in general and American culture in particular. The "American Dream" had an enormous impact in Norway, especially since the Marshall Fund had helped rebuild the country after the war. Bottles symbolized this culture for good and for bad. In 1963, a new generation of youth enthusiastically stated that "things go better with Coke!"[34] At the same time, broken glass from their empty bottles littered the roadside.

The Norwegian Ministry of Social Affairs received many complaints about disposable bottles starting in the early 1960s. Their first impulse was to ban these bottles, but they soon realized that it would be complex and difficult to achieve this in practice. Initially, the situation was resolved when the ministry negotiated a voluntary agreement between bottlers, brewers, and the Wine Monopoly on avoiding the use of disposable containers in 1965.[35] The Norwegian Brewery Association (Den Norske Bryggeriforening) later stated that they agreed to this because they already had enough trouble with the temperance movement and did not want to antagonize the environmentalists as well.[36] Thanks to the agreement, the stubby bottle disappeared within a few years. Still, the discussion over the stubby—and nonreturnable containers in general—signaled that the common interests of the brewers and bottles were dissolving. The bottle-driven cooperation between the members of the Brewery Association could no longer be taken for granted.

Small technological changes in bottling decisively altered the bottlers' frail balance between cooperation and competition. When two Norwegian bottlers introduced screw caps to Norway in 1970, the industry unity again came under pressure. Nora Fabrikker and Jarlsberg Mineralvann originally sold soda in twenty-centiliter bottles because this size acted as a single serving, consumable before the soda went flat. Screw caps made larger, resealable bottles more practical, so these two companies wanted to supersize their bottles to hold one liter. The Brewery Association at first convinced the two companies to not use nonstandard one-liter bottles, but this agreement did not last long when the temptation of increased sales grew too big. Soon both bottlers were selling one-liter bottles everywhere, and sales skyrocketed. The consumer demand for larger bottles was clearly there.

The thick, heavy one-liter glass bottles were expensive to purchase, so Nora and Jarlsberg wanted the bottles to carry a deposit. However, these bottles did not fit in the infrastructure set up to handle the reception of empty bottles in grocery stores, which is a part of the story that we will explore in the next chapter. When the grocers protested against these bottles by threatening to stop selling them, Jarlsberg instead targeted small convenience stores and gas stations for distributing the one-liter sodas. And the empty bottles ended up in the grocery stores after all, since this was where most consumers brought their empty bottles.[37]

Environmental Protection or Protectionism?

The efforts of grocers and bottlers to take control of the empty-bottle situation quickly escalated to a political level. After the introduction of the one-liter bottle, the Ministry of Social Affairs received demands for the regulation of empty bottles. This time not only littering but also the handling of the empty bottles had become a problem. Furthermore, international pressure for using disposable containers increased. In 1969, the Nordic Council, a Pan-Scandinavian forum for promoting common Nordic interests, proposed a joint Nordic stance on disposable containers that would allow use of these containers all over Scandinavia. They saw free trade as the most prominent advantage of entering into such an agreement.[38] The Norwegian government, however, did not agree with the Nordic Council. On the contrary, the combination of the problems in the voluntary deposit system, international pressure, and growing environmental consciousness led to a new government-sponsored proposal for disposable-container legislation.

Countries all over the world introduced environmental conservation in the early 1970s. Public and political attention was directed toward environmental issues. For example, the Council of Europe declared 1970 to be the European Year of Protecting the Environment.[39] It was also the year of the first International Earth Day, an American celebration of the vernal equinox, an observance adopted by the United Nations as an environmental-awareness-raising event in 1971. A slew of environmental laws passed in the early 1970s indicated a growing political interest in addressing some of the problems caused by consumerism and industrialization. The United States introduced the Clean Air Act, the Clean Water Act, and the Endangered Species Act, as well as created the Environmental Protection Agency just within a few years of the first Earth Day.

Environmentalism in the 1970s first and foremost arose as a reaction to industrial society, and only later turned its attention to consumer society.[40] Norwegian environmentalism—and Norwegian environmental laws—have their main origin in the issue of pollution from energy-intensive industries, particularly from places like the aluminum plant in Årdal.[41] The Norwegian public long accepted industrial pollution as a necessary part of prosperity, as part of the technological progress story that dominated post–World War II Norway. During the 1960s, however, as the problems caused by this pollution became increasingly visible, resistance to industrialization grew. The nature conservation movement attempted to protect nature from industrial development. Organizations like the Norwegian Society for the Conservation of Nature (Naturvernforbundet) fought to protect nature as an aesthetic resource. The protests against the damming of the Mardalsfossen waterfall in 1970 were the beginning of a large public environmentalist movement spearheaded by public figures such as Arne Næss and Sigmund Kvaløy, who went on to become international leaders in the deep ecology movement.[42]

The Nature Conservation Act of 1970 was the most prominent of the Norwegian environmental laws passed at this time, and it reflected contemporary environmental concerns. It primarily targeted natural conservation rather than environmental protection as such. In this context, the environment was not an abstract, global entity, but rather a local, landscape-centric environment that the Nature Conservation Act intended to protect. Empty bottles had some room in this debate, but only as part of a general opposition to littering and visual pollution. Paragraph 16 of this act prohibited the disposal of waste in nature in a manner that marred or harmed the natural environment.[43] The law did not specifically mention bottles, yet they became a significant issue within the environmental discourse.

THE DISPOSABLE CONTAINER ACT

Even in the early 1970s, Norwegian businesses frequently framed their activities in environmental terms, but such organizations generally had a more complex set of motivations. The first Norwegian beverage container legislation reveals how these conflicting interests both clashed and complemented each other in the creation of such laws. The act "enabling the prohibition of certain kinds of disposable containers in the marketing of consumables" (hereafter called the Disposable Container Act) passed in 1970 and gave the government the right to ban disposable containers (but the act itself did not actually ban the containers).[44] The act specifically stated that only disposable bottles were to be banned, not cans, since (in a rather ridiculous claim) cans would naturally degrade "over a somewhat extended period."[45] On the surface, the Disposable Container Act thus seemed to be an anti-littering act, with the aim of preventing the disposal of empty bottles in the landscape, but the Storting discussion raised doubts about such an interpretation.

When presenting the act for the Storting, Alfred Henningsen—the chairman of the Storting committee that considered the proposed act—stated that the Ministry of Social Affairs had drafted the act primarily out of environmental concerns. However, Henningsen and the committee could not recommend approval of the act, arguing that since it only covered disposable glass bottles, it would not be able to target the real littering problem at all. In his view, other forms of disposable packaging, such as plastic shopping bags, represented a larger littering problem. Norwegian breweries did not even officially use disposable bottles in 1970, so littering could not be the real reason for the law. He questioned the motives behind the act, as well as the process by which it had been created. First, as far as he could tell, the organizations who requested the act were all representatives of the bottling industry. Second, he wondered why the Ministry of Industry and the Industrial Council had not been consulted. Third, and strangest of all, he questioned how they could make a supposedly environmental act without coordinating it with the Ministry of Local Affairs' new proposal for a Nature Conservation Act. Henningsen really did not know what to make of the law. How could it be an environmental law if it neither covered the most common disposable containers nor considered the Nature Conservation Act? How could it be a law regulating industry if industrial interests were not consulted?[46] In fact, several newspaper articles argued that it was the bottlers themselves who asked for this act, primarily to prevent international competition.[47]

The conservative government defended the act. Jo Benkow, who argued for the act on behalf of the conservative bloc that held the majority in government, argued that just because we did not have a widespread problem with littering today did not mean that we shouldn't have the means to solve this problem should it arise. This was the main purpose of the Disposable Bottle Act as he saw it: "We have not yet had to deal with the overwhelming amount of garbage following in the wake of affluence."[48] In his opinion, disposable bottles would lead the way in creating a throwaway society. Benkow's defense of the act was convincing enough—or the conservative bloc strong enough—to win a narrow victory in the vote with forty-one in favor versus thirty-nine against.

BUSINESS PROPONENTS AND OPPONENTS

The 1970 Disposable Container Act allowed the prohibition of disposable bottles, but did not address cans. Soon the government would be forced to consider an expanded act covering cans. During the planning on this expanded act, different industrial interest groups flooded the government with requests, statements, and arguments. The beverage industry was the most prominent of these groups. However, this industry was split on whether it wanted to allow disposable containers. While they sometimes used environmental arguments, the business groups picked positions in the debate based on their own economic interests. Large bottlers stood to gain from disposable containers, while the smaller, mostly local companies clearly preferred to maintain the current deposit system.

The National Soda Bottler Trade Association (Mineralvannfabrikantenes Landsforening), which organized many small bottlers, stated in a letter to the Ministry of Industry that it was important to expand the law to forbid the use of cans and disposable bottles as soon as possible, before the "strong economic interest groups" that wanted to introduce these containers gained momentum.[49] These interest groups included both large Norwegian breweries as well as international competition. In addition, the association saw cans as the largest threat to the environment and to the industry in the future, arguing that "if the development goes as fast as in the U.S.A., the picture will look very dark from an environmental point of view."[50] They frequently used the situation in the American beverage industry as a negative comparison, as something to avoid. In addition to the ostensibly environmental arguments, industrial concerns played a central role in the debate. The association argued that a ban on disposable bottles that did not include cans would strike small bottlers harder than large ones. First, the high cost of installing the canning machinery prevented small companies from making this investment. Second, the increased competition that would come from foreign, "gigantic" bottlers would outcompete domestic beers. Third, the increased reach of the larger bottlers' distribution chains would increase the competition for local bottlers even more. The outcome of the act could thus significantly change the conditions for the Norwegian beverage industry. As a result of the pressure from the small bottlers, the government started working on an expanded Disposable Container Act that would include both bottles and cans and levy a deposit on these containers. It would take four years and much angst to put together new legislation.

On the other hand, the similarly named National Soda Industry Trade Association (Mineralvannsindustriens Landslag), which organized the larger bottlers, had no problems with allowing cans and disposable bottles. In fact, several member bottlers had begun making plans for and investments in large canning and bottling facilities. For instance, Noblikk-Sannem AS, a major producer of tin cans, worked with Coca-Cola to build a canning facility that could supply all of Scandinavia with canned Coke. A ban on cans would "amputate their future," as they put it. Instead of a ban, they wanted an informational campaign to educate Norwegian consumers: "We believe that it is only a minority of our people who have the bad habit of throwing cans, bottles, and cardboard boxes in nature, since the majority of Norwegian women and men still are such great nature lovers that they understand the importance of keeping clean the treasure that we call Norwegian nature."[51] Noblikk-Sannem also reminded the Ministry of Social Affairs that the company employed 612 workers. The larger bottlers and their unions frequently invoked labor interests in the disposable-bottle controversy.

The correspondence between the two bottlers' associations, the Ministry of Social Affairs, and the Ministry of Industry reveals that Norwegian brewers and bottlers no longer had a common interest in maintaining the existing system with deposits on reusable bottles. Some of them felt more threatened by international competition than others. This division was particularly visible in the relationship between the two trade associations for bottlers and brewers.

Other business groups also protested against the expanded act. For example, the Norwegian Business Community Association (Norges Handelsstands Forening) fully supported the environmental intentions behind the act, but would not approve of the suggested beverage container deposit. The association believed that the practical problems and high costs of such a system would stop its members from selling and marketing goods included in the law.[52]

The aluminum and glass industry clearly had a stake in the new legislation. Norsk Hydro, one of the largest Norwegian aluminum producers, asserted that it wanted to avoid any restrictions on free trade.[53] SkanAluminium, the Scandinavian aluminum industry trade organization, stated that a ban on aluminum containers would have a significant impact on the Norwegian aluminum industry, which was the largest in Europe.[54] Elkem Aluminium AS, which had established a cooperative relationship with Alcoa on producing Alcoa's patented "easy-open" cans, stated that a can ban would "have a direct impact on Norway's export potential for crude aluminum."[55] Further, glass producers like Moss Glass Works flatly refused to support the law because of the proposed deposit system and ban on certain material types.[56] Disposable bottles represented a new and promising market for them. On the other hand, cans represented a direct threat to Moss Glass Works, the only producer of glass packaging in Norway, employing 680 persons and supplying about 85 percent of the Norwegian demand for glass packaging.[57]

The Norwegian Industrial Association (Industriforbundet) had a subgroup called the Packaging Forum, which was one of the most outspoken groups in the discussion. The forum compiled a list of arguments against the act that it presented to the Ministry of Industry.[58] First, it asserted a direct relationship between modern disposable packaging and the modern industrial welfare society. A prohibition of disposable packaging would thus threaten "the standard of living that we enjoy today." In doing this, it basically turned the problem upside down. In the forum's view, disposable packaging was not a consequence of affluence—it was its cause. Second, it argued that disposable packaging did not in fact play a significant role in causing environmental problems. Third, a ban on disposable containers would not reduce the total amount of waste—just change its composition. Fourth, the packaging industry was in its view not the right group to target, as it was people, and not the industry, that caused littering. The littering problem could be solved only by targeting the litterbugs through enforcing current laws. Fifth, a return system for disposable packaging would in practice turn every family into a "mini sanitation facility" and every store into a "mini sanitation department." This would lead to an immense bacterial accumulation; an enormous (and with the handwritten addition "and extremely costly") collection system; and a gigantic control system encompassing grocers, authorities, and all users of packaging. Finally, a ban on disposable packaging would lead to employment loss in the Norwegian packaging and materials industries. The Packaging Forum thus used the same line of arguments that American industry-sponsored organizations like Keep America Beautiful had developed. People—and not industry—were to blame for littering, and they could be targeted only through information campaigns aiming to change people's mentalities.[59]

Not surprisingly, the Ministry of Industry sided with the majority of the business community—in other words, the large companies—in opposing the expanded Disposable Container Act. In a 1971 meeting with representatives from the Industrial Association and the glass and can packaging industry, the ministry emphasized the practical difficulties of enforcing such an act.[60] They sent a letter to the Ministry of Social Affairs describing the proposed deposit system as "neither realistic nor feasible."[61]

So while many of the actors in this controversy argued for environmental protection, commercial motives clearly underlay their opinions. However, the industry groups did not have common interests. They were split in regard to national versus international and local versus regional distribution—something that primarily translated into large versus small bottlers. In addition, labor interests and unions used both economic and environmental arguments for what they were worth to protect their members. When they made claims to environmental concerns in this discussion, they did it as a way of protecting their interests while arguing for the common good at the same time.

ECONOMIC SOLUTIONS TO ENVIRONMENTAL PROBLEMS

In Norway, the Ministry of Finance became involved in setting up a government-mandated bottle deposit system. To understand the origins of bottle deposits as a form of environmental regulation rather than an economically motivated bottle buyback, we need to look into the architects behind the regulations. The Ministry of Finance secretary general Eivind Erichsen was trained as a social economist and served as one of the two Norwegian administrators of the Marshall Plan and as secretary general in the Ministry of Finance. In 1971 Erichsen wrote an article titled "Do We Have to Choose between Environmental Protection and Economic Growth?"[62] In this article, he argued that our environmental problems stemmed from the postwar economic growth. The main problems were pollution of air, soil, and water; waste generation; and a lack of access to natural experiences for city dwellers. In Erichsen's view, economic growth had to be directed in a way that was compatible with environmental protection. He wanted to use the market mechanism to achieve this, particularly by employing taxes and fees on waste to encourage industry to develop more environmentally friendly products and processes. We can see this as a direct precursor to the idea of sustainable development, which will be further discussed in Chapter 6.

Erichsen used his position in the Ministry of Finance to implement his plans for environmental taxes. In 1972 he suggested a small environmental "package" consisting of fees on canned beer and soft drinks and on plastic grocery bags and increased bottle deposits. This proposal was later expanded to include fees on scooters and motorcycles, motored pleasure boats, and detergents containing phosphate. While these fees would increase the state's income, Erichsen argued for the importance of giving the taxes an environmental profile. Erichsen's tax package did not have as its purpose to "solve the great environmental problems," but instead targeted issues that were clearly set out and where one could get quick results.[63] The act was intended to regulate and control the use of "unnecessary" and "wasteful" products.[64] While this on the one hand was a reflection of the polluter-pays principle—arguing for the producer's responsibility for environmentally damaging products—it was also influenced by a moral, ascetic approach to consumption as something that should be controlled and regulated. In other words, he argued for taxes that would encourage everyday environmentalism.

Taxes became one of Norway's key approaches to environmental issues. The Product Control Act of 1975 was one of the first to be passed by the new Ministry of the Environment.[65] It was a result of negotiations between three ministries—environment, finance, and industry—demonstrating the complex coalitions involved in environmental legislation. These three ministries took the position that financial incentives like taxes and deposits were the best way to motivate consumers and businesses alike to stop polluting. The main advantage of these incentives was that they would allow consumers to keep their free-market choices, while at the same time encouraging companies to introduce new and more environmentally friendly products. Deposits would stimulate the recycling of materials and prevent hazardous products from being thrown away. Norwegian environmental policy clearly found its place and identity in the early 1970s within an economic framework. Environmental problems were defined as an economic problem that could be solved through taxes and financial incentives. The bottle deposit system became a frequently mentioned model for this framework.

Within this framework of environmental regulation, the Ministry of Finance asked Norwegian bottlers and brewers in 1974 to voluntarily increase the deposit to from thirty to fifty øre for containers up to fifty centiliters and from seventy øre to one krone for larger containers, as the old deposits "did not seem to be sufficient to prevent empty beer and soda bottles from still littering nature."[66] The return rate increased to 99 percent following the deposit increase in February 1974. However, not all bottlers followed the Ministry of Finance's request, which created problems in the common-bottle-stock system that the brewers maintained. Unless all bottlers used the same deposit values for the bottles, the system would lose money, since bottles with a thirty-øre deposit could receive a fifty-øre refund.

Because the bottlers failed to institute an increased deposit, the Ministry of Finance drafted a law called the "Temporary Act on Deposits on Packaging for Beer, Soda, and Other Drinks" (Bottle Deposit Act) to mandate the deposit value.[67] Storting passed this act in May 1974. Through this move the Ministry of Finance made all of the preceding discussion on the Disposable Container Act and its expanded version moot. The act specifically aimed to use the deposit-return solution that the brewery industry had successfully implemented throughout the twentieth century "to prevent empty bottles from littering nature."[68] The tried and true returnable bottle of the beverage industry was now put to a new and ostensibly environmental goal.

The 1974 Bottle Deposit Regulations, which implemented the 1974 temporary act, set a minimum refundable deposit on all beer, soft drink, and other beverage containers (both returnable and nonreturnable) of fifty øre for sizes of fifty centiliter or less and one krone for larger ones in order to ensure a high rate of recycling or reuse.[69] These regulations required all bottlers and brewers by law to comply with the deposit regulations for beer and soda packaging, beginning July 3, 1974.[70] The most notable difference from the previous acts was that disposable containers, including cans, now were subject to an eighty-øre tax on top of the deposit. This made them so expensive compared with refillable containers that in practice they were banned. It was under the auspices of this temporary act that the government was able to institute a specific environmental tax on cans and other disposable containers, effectively keeping them out of the market.

Small Things, Big Problems

As we have seen in this chapter, bottles represented many things to many different actor groups in the 1960s and 1970s. All these small problems added up to become a large problem, reaching deep into Norwegian industry, consumer society, and the emerging environmental policy makers. The old deposit system depended on common interest and a cooperative attitude among the bottlers and brewers. When common interest grew smaller and smaller, cooperation grew more and more uneasy. Additional groups became involved in the discussion on bottles, redefining what the bottle problem was. Policy makers saw bottles both as a littering problem and as a source of income. For them, bottles represented something that should be regulated and controlled. At the same time, consumers wanted convenience, affordable goods, and—occasionally—environmentally friendly solutions. While some actors were more powerful than others, none were strong enough to completely dominate the discussion. The environmental aspects of the bottle deposit system became a rhetorical arena where the business actors faced each other, switching between environmental and economic arguments.

In the end, a high tax on disposable containers—both bottles and cans—ensured that reusable bottles would continue to dominate the Norwegian market. The use of economic environmental incentives like taxes and fees took shape in late 1960s and early 1970s, modeled on the relative success of the voluntary bottle deposit systems. This was similar to what happened in the U.S. states that chose mandatory deposits as the solution to the one-way beverage container problem. During these years of political discussions about empty bottles as business and environmental problems, however, other groups had to face the practical consequences of the bottle returns.

The controversies over beverage container taxes in the early 1970s in Norway shows us how a well-established system, such as the industry-run glass bottle return system, can erode as the interests of the actors maintaining them shift and as new technologies and new materials enter into the picture. In the case of the beverage industry, the technological and organizational choices that defined the system privileged certain actors over others. In the following chapters we will see exactly how the technologies and legal structures that shaped and made possible modern beverage container recycling came to be, and how environmental concerns increasingly became a key parameter in these systems. More important, such environmental arguments were invoked by actors on all sides in the beverage container recycling controversies.

Chapter 3 Creating Bottle Infrastructures

ON THE NIGHT OF JANUARY 2, 1972, four men gathered around a hole in the wall in a small grocery store in Oslo, Norway. In the backroom behind the steel-framed, rectangular hole, they had just installed the prototype of a remarkable machine soon to be found in stores all over Europe. Next to it stood a shopping cart full of empty glass bottles that the men fed this machine. A conveyor belt transported the bottles through a hole and into the backroom. After all the bottles had passed through the hole, one of the men pressed a button on the machine. A printed receipt then came out of a slot next to the button. When they saw that the number printed on the receipt matched the number of bottles they had put into the shopping cart, they were all in high spirits—but not for the same reasons.

In some ways they saw two different machines when they looked at the hole in the wall. The group's two grocers, Aage Fremstad and his mentor, Kolbjørn Jacobsen, saw the machine as a technological solution to the increasing number of empty bottles that piled up in the backrooms of their stores. For them it was an infrastructure machine that would reduce labor costs and increase their control of bottle deposits and refunds. It could clean up the messiness of bottles in their backrooms. The other two men, entrepreneurs Petter and Tore Planke, on the other hand, saw a business opportunity in solving the grocers' bottle problem.[1]

At thirty-five years old, Petter Planke was the older of the two brothers and an experienced salesman with a successful career in the grocery industry. Yet he had a long-standing dream of creating and building his own business. If the machine worked as they hoped and if he could sell it to storeowners, he might realize his dream. Tore Planke was seven years younger than his brother and had completed his engineering degree at the Norwegian Institute of

Technology (NTH) just a few years earlier. He had been very busy lately. While designing and building the prototype reverse vending machine (RVM), he held a daytime job at the Engineering Research Foundation at the Norwegian Institute of Technology (SINTEF), where he worked on developing the world's first fully automated navigation systems for supertankers.[2] He also had a family to look after. The night before they installed the machine in Fremstad's store, his wife gave birth to their first child. This did not stop him from spending the following day with his brother, first tearing down a wall in the grocery store and then installing and testing the machine. While it cannot have been an easy decision, his choice of priorities illustrates how important the machine was to him, and perhaps also the amount of work expected of entrepreneurs. When the Planke brothers looked at the machine, they saw the result of their creativity, knowledge, and hard work. They saw a machine that they believed had the potential to make a difference and they were willing to quit their day jobs for it. They had addressed the grocers' bottle problem, and saw it as their key to new, meaningful, and exciting careers for them both.

After testing the machine and confirming that it counted all the bottles correctly and that there were no mechanical problems, Petter Planke got out a bottle of champagne to celebrate. He was already planning how to build up a business to market and sell this machine to grocers like Fremstad and Jacobsen, who indeed became the Planke brothers' first two customers. They were happy to purchase a machine that had been designed to their exact specifications. However, by applying advanced technology to the mundane problem of bottle handling, Tore Planke had created a machine that had the potential to extend far beyond the backrooms of Norwegian grocery stores.

This chapter takes the grocers' bottle problem as its starting point and follows the start-up of Tomra Systems in the backrooms of grocery stores in 1971. As we saw in the preceding chapter, the interbottler agreements behind the Norwegian bottle-recycling system had grown increasingly fragile, and the grocers who did the actual handling of returned bottles found the existing solutions to the problem inadequate. When the grocers approached Petter Planke with this problem, he and his brother were able to convert the specific user demands into functional engineering solutions that could be embedded into a machine. At this point, the economic and logistical concerns of the grocers dominated the machine's development and distribution. The specific problems it was designed to solve will be analyzed in detail, as will be the technological and organizational solutions it replaced. This will show us the historical origins of an RVM-centered recycling system.

We ended the previous chapter by looking at the creation of the first bottle recycling laws in Norway. Even before these political discussions about bottles and littering took place in the parliament, others were dealing with the tangible problems of bottles. The returned bottles piled up in the backrooms of grocery stores, presenting a large amount of work for Norwegian grocers. Both before and after the Bottle Deposit Act went into effect, grocers served as the front end of the deposit systems. Even though this way of returning and reusing bottles made perfect sense for the bottlers, it represented no small amount of work for the grocers. The average Norwegian grocery store in the 1960s was—and still is—rather small.[3] Depending on how large the staff was, stores received empty bottles either in a separate room in the back of the store or right next to the checkout counter. In both cases the store staff had to receive and count the bottles that the customers returned, calculate the total amount to be refunded according to the deposit values of the different bottles, and then pay the refund to the customer. The bottles then had to be moved out of the way, so that there was room for next customer to return his or her empty bottles as well as make store purchases. After the store closed, all the bottles were counted again and placed in crates to be returned to the bottler the next time the delivery truck came. Needless to say, grocers faced several logistical challenges in this operation. Bottles took up a lot of space and they were heavy. In addition, the bottles were not necessarily clean when returned. Most contained small amounts of soda or beer, which becomes rather smelly after a while. This made the work situation unpleasant and unhygienic for the staff, as well as making the store environment less attractive for consumers.

Handling empty bottles thus represented both hard and unpleasant physical work and a time-consuming calculating process for the staff in grocery stores. On Saturdays—the main shopping day—the line of people waiting to return their bottles could go through the entire store. At times like this, the store needed to have dedicated people handle the bottle returns. On other days, the regular staff had to take care of this task, which constantly interrupted them from their other responsibilities. Meanwhile, the backroom was overflowing with empty and smelly bottles. It is no wonder that grocers were not pleased with this situation.

While there were existing technologies to handle the return of empty bottles, these did not work satisfactorily for grocers, especially as new bottle types like the one-liter bottle entered the market. In addition, grocers had not been able to negotiate organizational solutions to the bottle problem. They had been relatively silent during the political discussion about the bottle deposits, yet they were directly affected by it. Although Tomra would later enter into political and environmental arenas, it was in fact the grocers' bottle problem that was the starting point of the company's reverse vending machine.

Figure 1. Grocer stacking crates of empty glass bottles, 1965. *Courtesy of Oslo Handelsstands Forening.*

The Rise of Automatic Vending

Throughout the 1950s and 1960s, various forms of vending machines and automats became increasingly popular in grocery stores.[4] The Oslo Grocers' Association's journal *Kjøpmannsnytt* frequently featured articles on automatic vending, often stating how these machines were common abroad but not much used in Norway yet. However, "with the increased use, the automatic culture will follow—the machines will be left in peace and instead become a natural part of human existence."[5] Through statements like this, grocers increasingly portrayed vending machines as an integrated part of a modern store. In the supermarkets that were appearing across Norway, self-service became the norm. The automatic vending machines were not only for selling things; some simple machines could accept empty bottles—reverse vending machines.[6]

To attract customers, distinguish his store from competitors, and of course help with the bottle handling, the grocer Per Fremstad installed Norway's first machine that customers could use to return their empty bottles in his store, PF-Hallen, in Oslo in 1957. The machine was made by the Swedish company Wicanders, which primarily made bottle caps and machines for capping. "The Bottle Crusher," as Fremstad called it, received bottles through a hole and used hydraulics to push them onto a table in the backroom. This system was not very sensitive, hence the name. It was also very noisy, leading Per Fremstad's son Aage to send a letter of complaint to the producer. "We cannot change the construction in any way to measurably lessen the din," Wicanders replied. Instead, it encouraged Fremstad to install sound-dampening plates around the machine.[7] Wicanders's decision to not improve its machine indicates that it did not give much priority to something that it saw as outside its main business focus.

The Wicanders machine presented customers with a new way of returning their empty bottles and allowed the grocer to delegate bottle-handling tasks to the machine. For each bottle, the consumer would receive a red or blue token, according to the refund value. These tokens were given to the cashier, who counted them by hand and gave the money back. Per Fremstad used the machine in the marketing material for the store, and customers thought it was exciting to send the bottles through a hole in the wall instead of giving them to the store personnel. The novelty effect of performing an everyday task in a new and different way probably contributed to attracting new customers to the store. As such, the machine not only performed a physical task, but also did work on a symbolic level. Automatic bottle return machines were a visible indicator of a modern, forward-looking grocery store. The practical effects, on the other hand, were limited by the inconvenience of counting tokens, the constant noise, the frequent broken bottles, and the high maintenance costs.

When Per Fremstad died in 1957, less than a year after opening PF-Hallen, his son Aage took over the store. Aage Fremstad had completed a one-year business education with good grades. Although he wanted to study more, his plans were interrupted by the German occupation of Norway during World War II. After the war, he felt that it would take too much time to continue his studies, so he instead took a job as a salesman for a manufacturing firm. Through his job, in which he sold large industrial equipment such as excavators, Fremstad was exposed to an international engineering environment, traveling to both England and the United States. While not an engineer himself, he became familiar with the discipline's general knowledge and way of looking at problems. This would prove to be valuable later in his career. Fremstad admits that he "did not know anything about retailing" when he started, but he was helped out by his father's friends and connections in the grocery industry.[8]

SEARCHING FOR TECHNOLOGICAL ALTERNATIVES

When Aage Fremstad began working full time as a grocer, he soon replaced the bottle machine that Wicanders refused to improve with a slightly more advanced model made by the Norwegian company Arthur Tveitan AS in 1962. Both machines were distributed by the Swedish company Hugin. Tveitan's machine was designed to solve some of the problems with the Swedish machine. Most important, it was not as brutal with the bottles and was a lot quieter. Arthur Tveitan AS was a small company, founded in 1936, specializing in ticket-issuing machines, cash registers, scales, and other technical equipment for retail stores. Its first bottle machine, used for returnable milk bottles, launched in the early 1950s.[9] Every bottle gave the customer one deposit slip. Later machines used marbles instead, with different colors, depending on the bottle types and deposit value.

Arthur Tveitan's son Fredrik did not have any technical education; yet he designed Tveitan's bottle machine and continued developing the technology for several years. He built a simple and affordable machine based on mechanical recognition of bottles. Tveitan's machine can be seen in Figure 2 on either side of the Tomra prototype RVM. The machine had one opening for each different kind of bottle, which amounted to six or eight different holes, and several conveyor belts that pushed the bottles out onto a table on the rear side of the machine. The customer would insert a bottle in one of the holes and pull a handle to push the bottle through, receiving in return a coupon with a hole punched in it. If the customer had twenty bottles, he or she had to do that twenty times and get back twenty coupons, and they all had to be counted by hand at the cash register. According to grocers, counting all these coupons was too time consuming. The machine could also easily be tricked, as it identified bottles by mechanically measuring their height. A detergent or ketchup bottle of the right height would thus be identified as a deposit bottle, causing the grocer to lose money.[10]

Despite these small problems, most grocers thought this machine was a big improvement over handling the bottles manually. It was also very affordable. This gave it a reasonably large user base among the grocers, including the Norwegian Consumer Co-operative (Norges Kooperative Landslag) stores.

Figure 2. Prototype Tomra I RVM in the middle with Tveitan bottle machine on both sides. The openings in Tveitan's machine are blocked, as the machine was no longer in use. Tveitan's machine had one opening for each bottle type, whereas the Tomra I had only one opening used for all bottles. Note the user instructions on the wall by Tomra's machine. The prototype had to be manually turned on and off by the user, and this confused many customers. Tomra's production models used an automatic sensor for this purpose instead. Courtesy of Tore Planke.

In 1968 there were 858 of these stores spread all over Norway.[11] While Tveitan mostly sold to the Co-op stores in the Oslo region, independent grocers also bought the machine. In addition, some sixty machines were exported to Sweden.[12] As late as 1990, 360 Norwegian stores still used Tveitan's machine.[13]

Grocers also tried other technical solutions to the bottle problem. One large Oslo store installed a conveyor belt leading from the tobacco and chocolate counter—where customers also returned bottles—to the backroom. Even though the bottles still had to be manually counted by the store personnel, the belt got the bottles out of the way. This setup cost about ten thousand kroner and worked well enough, according to the store manager. However, since the refund was paid in cash and the store did not require its customers to buy anything in order to get their refund, they received a lot more bottles than they sold. Obviously, the store manager saw this as a problem.[14] A large bottle surplus made handling and storage even more challenging for grocery stores.

While the bottle return machines made returns more convenient for the customers, they could actually create more work for the grocers. Still, technologies such as these demonstrate the grocers' continuing efforts to delegate the task of bottle returns to machines.

CONTROL, COUNTING, MAINTENANCE, AND FLEXIBILITY

Even with machines in place, four problems still frustrated the grocers. The first problem was control. Some grocers claimed they had at least a 15 percent loss on bottle returns. The existing machines could not be relied on to always recognize valid bottles. Customers could claim they had delivered ten bottles when they had delivered only five, and demand the rest of the money from the grocer. As previously mentioned, Tveitan's machine could also be tricked into recognizing any bottle of the right height as a deposit bottle. The old system was based on honesty, which obviously did not work very well. Empty bottles were a good source of extra pocket money for kids, so many of them were tempted by the opportunity to make a few extra kroner.[15] For the grocers, however, these small sums added up to a large problem. The grocers who had bottle machines could also run into problems with internal theft. Dishonest employees could insert the same bottles several times and claim a refund.

The second problem was the counting of tokens and coupons. The machines could not sum up the total amount of bottles returned, but rather gave the consumers various forms of tokens in place of the empty bottles. This required cashiers to spend an inordinately large portion of their time counting coupons. This was amplified by the fact that most people came with their empty bottles on Friday and Saturday afternoon, which would first create long lines in front of the bottle return machines and later at the checkout counter where the cashier had to count the tokens. Stores often hired extra personnel to help them during these periods. The opening hours were generally shorter than they are today, so customers also had a shorter time window in which to return their bottles and shop for groceries.

The third problem was the high maintenance costs. Sticky soda or beer leftovers unerringly found their way into sensitive mechanical parts, causing the machine to jam and stop.[16] Straws or loose bottle caps could also cause jams. Posters next to the machines attempted to educate customers to clean bottles properly before returning them, but the effect of these information campaigns was, and still is, limited. This was one of the reasons why many grocers decided not to purchase bottle return machines, but instead continue their manual bottle handling.

The last problem was the introduction of new bottles that the machines had to handle. The one-liter beer bottle that put the deposit system under pressure around 1970 simply did not fit in Tveitan's machine. The machine had been custom built to Norwegian specifications, but the bottle recognition could not be reprogrammed to accept new bottle types. The one-liter bottle thus had to be handled manually even if the store had a machine. Because of the thick glass required of these large bottles, they were also very heavy, making the work of handling them physically demanding. All in all, the machines were not sufficiently adapted to the challenging material and social spaces in which they operated. They were not flexible enough to cope with changes in their use. Because the old machines were created to solve the problems of the old bottle deposit system—with few bottle types and a united beverage industry working to maintain the deposit system—they failed to perform their task in the new system that arose in the late 1960s. They did not fulfill their promise of relieving grocers of hard physical labor or time-consuming counting.

Looking for a Solution: The Empty Bottle Committee

Many grocers grew tired of doing this hard and unpaid work of sorting bottles for the breweries. In 1967—years before the disposable-container discussion escalated to a legislative level—they organized an "empty-bottle committee," as they called it, to find a solution to the bottle problem. The committee consisted of members from bottlers; breweries; several large retailing chains, including the Norwegian Consumer Co-operative, and two grocers' associations. Kolbjørn Jacobsen represented the national association and Aage Fremstad the Oslo association. Fremstad had been invited to join the committee for two reasons. The first was because his father was well known and highly respected among the Oslo grocers. The second, and more important, reason was Fremstad's background as a salesman of technical products. Fremstad later recalled how the other members of the committee would just call him "the engineer."[17] The committee hoped his knowledge of technology and engineers would help them find reliable and future-proof solutions to their problems.

Initially the committee was concerned mainly with organizational approaches to bottle deposits. The grocers had suggested already in the early 1960s that disposable containers would be more rational for them and asked the Norwegian government to change its policy and permit the use of disposable containers.[18] At the time, the government did not want to change the current deposit policy. This made it clear to the grocers that there was no way to avoid a deposit system. Empty bottles would continue to be part of grocers' lives.

The empty-bottle committee met regularly from 1967 to 1972. It proposed the introduction of a differential deposit, whereby the bottler pays the grocer more for a returned bottle than the grocer paid as a deposit for the full bottle. This would ensure that the grocers were compensated for their work. In this system, suggested in early 1971, the deposit for large bottles, including the new one-liter bottle, would be set at seventy øre from producer to retailer and sixty øre from retailer to consumer. Correspondingly, the deposit for small bottles would be thirty-five and thirty øre. In their attempt to convince the beverage industry to change the deposit system, the grocers used financial and logistical arguments. For instance, the committee stated that a higher deposit would make more bottles come back to the breweries sooner. The bottle committee also raised concerns about the environment and brand visibility of littering, stating that increased deposit values would "have environmental advantages. There will be fewer one-liter bottles left in nature to create negative goodwill for the breweries."[19] By doing this, the grocers latched on to the emerging debate on littering and environment described in the previous chapter.

However, the breweries did not want a deposit fee that was higher than the price of a new bottle, which was forty-five øre. Such an arrangement would undermine the economics of the current deposit system. If new bottles were cheaper than the deposit value, it was theoretically possible to purchase new bottles and then return them in the grocery store for a profit. Furthermore, if some bottlers decided to sell their bottles without a deposit, their bottles could still be returned in the stores. The bottlers who still had a deposit system would then end up paying for these bottles. The proposal was ultimately denied by the Norwegian Price Directorate because there was a current freeze on prices of consumer goods.

The 1974 Bottle Deposit Regulations required bottle deposits, but before then, the lack of a mandatory deposit meant that the grocers could not get the bottlers to agree to their proposition. There were too many variables with bottlers setting different deposits and varying bottle sizes, and even some stores setting their own deposit amounts for containers. In the end, many grocers were so fed up by this process that they stopped sorting bottles altogether.[20] After four years without any results, the bottle committee decided that further negotiations were not realistic.[21]

When the grocers' negotiations with the beverage industry did not succeed, the committee started looking for a technical solution to their problem instead. If they could neither avoid the deposit system nor get paid for their work, the only sensible solution for them was to find ways to make the work easier and more practical. As we have seen, several stores experimented with different technical solutions to the bottle problem, but it was expensive and inefficient for individual stores to do this. The performance of the existing solutions was also limited.

In 1971, the empty-bottle committee decided to form a technical subcommittee with representatives from grocers, bottlers, and breweries to survey the national and international market in order to find a new and better machine. The committee found no existing automatic solution that covered their needs.[22] They asked the current machine suppliers—Hugin in Sweden and Arthur Tveitan AS—to consider improvements to their existing machines.[23] However, these firms found it too complicated or of little interest; they did not see bottle returns as a worthwhile business activity.

The solution that the grocers were looking for came from outside the established industry.[24] One day Fremstad, who was on the technical committee, voiced his frustrations to Petter Planke, saying "We are drowning in empty bottles and we don't have any way of handling the new bottle."[25] At the time, Planke was a salesman of automated machines for applying price tags for the Danish company Antonson Avery. These machines had been a great help to Fremstad earlier, and now he hoped that Antonson Avery could make an equally useful automated machine for bottle returns. Planke passed this request on to his employer, who judged it outside the company's core business area and thus not worth pursuing.

However, Petter Planke had been looking for a good opportunity to start his own company. Fremstad's bottle problem stuck in his mind as a chance to achieve this goal. While he did not have the necessary technical skills, his brother Tore was an engineer. Together they began discussing the possibility of developing a new bottle machine. They came from a family of entrepreneurs: their grandfather owned draper's shops and textile manufacturing firms in Berlin and Chicago and their father, Sverre, took over his father's leather goods company and often involved the children in the daily tasks.[26] This exposed the sons to the skills and values of running a business as well as to the problems and challenges that inevitably occur.

Petter Planke was born in 1936, the first child in the Planke family. He turned out to be resourceful, enterprising, and handy with tools. By the age of twenty-five, he had built two weekend cottages, owned his own house, was married and had two children. He earned a degree from Oslo Business College (Oslo Handelsgymnasium) and began working for Antonson Avery. After fourteen years at this firm, he was an experienced salesman with an extensive network of contacts in the grocery and retailing industry. Petter knew the world of grocers—how to talk to them and how to sell to them. During his experience of introducing self-adhesive labels to Norway, he learned some important lessons on salesmanship and on how people relate to new ideas, "no matter how good they are." His philosophy was that even if people agreed that something was a good product, they might still say that it's not for them. "We don't like to be early adopters," he concluded.[27]

Tore Planke was seven years younger and had many of the same qualities as his older brother. Unlike Petter, he pursued a university education, earning an engineering degree in Cybernetics at the Norwegian Institute of Technology (Norges Tekniske Høyskole [NTH]) in Trondheim from 1964 to 1970. His main teacher, Professor Jens G. Balchen, was one of the founders of engineering cybernetics in Norway. Until he retired in 1997, Balchen was an enthusiastic professor who trained generations of engineers who went on to have significant influence on the Norwegian technology sector, many of them starting their own companies. Tore wrote his thesis on "man-machine interaction in data control systems," mostly because he was dissatisfied with the "engineering solutions" that dominated the user interfaces in many of the projects he worked on as a student. This work had a direct influence on the Tomra bottle machine design.[28]

During his years at NTH, Tore acquired a solid training in what today is called *mechatronics*, a discipline combining mechanics and information technology (in other words, hardware and software) to integrate their functionality. While this traditional definition focuses on hardware, software, and mechanical components, Tore Planke and Tomra gradually learned to move the RVM out of these technical domains and into cultural and political spaces. Together the brothers were well qualified for taking on the task of building a better RVM.

Designing a Better Reverse Vending Machine

The grocers' explicitly stated problems were integral to the development of the first machine. Because of Petter's experience with the grocery industry, he functioned as an intermediary between the grocers and Tore Planke, translating user requirements into technical specifications. With his sales background, he was also able to market and sell the machine to grocery stores. The Planke brothers participated in a bottle committee meeting and requested exact specifications from grocers: What bottle types did they handle today? What would they need to handle in the future, both in Norway and abroad? Based on the conversations with the grocers, they broke the bottle problem down into four individual requirements: control, convenience, reliability, and flexibility.

First, the machine needed to recognize different bottle types and only give refunds for acceptable bottles, thereby controlling the flow of deposit money. Tore tried several different ideas for how to recognize bottles. Using a video camera seemed an ideal solution, but he found that the available technology was not good enough. Since bottles are transparent, it was very hard to get a decent image. The available camera lenses also introduced various optical errors and distortions. Correcting these errors was possible, but required more computing power than Tore could provide at this time. He then turned to a photocell detection system. Light sent out from one side of the bottle toward sensors on the other side could provide the machine with a numerical contour representation. After conducting some experiments, he went with this solution. His prototype recognized bottles based on their profile against a background of photocells. Considering that the existing RVM models all used mechanical recognition of bottles, this was a major technological step.

Second, the machine had to provide convenient bottle handling both for the consumers and grocers. Unlike the other models on the market that had separate openings for each bottle type, Tore's machine had one hole that would receive all bottle sizes and an electrical sensor that would determine the refund. It could identify up to eight different bottle types and their deposit values, using the photocell recognition technology. A conveyor belt transported the empty bottles to the back room. A printer would make one receipt with the total amount to be refunded by the cashier. This was a major improvement over the Tveitan machine, which at the time gave one receipt per bottle. It could also print a complete summary for the grocer at the end of the day.[29]

The reliability of the mechanical elements of the machine was just as crucial as the bottle recognition functions. As previously mentioned, the existing bottle return machines required frequent maintenance because of the high volume of bottles passing through them. Sticky soda leftovers quickly made complex or intricate mechanical components fail. Tore kept the movable parts to a minimum and attempted to coordinate the mechanical and electronic components as much as possible.

The most important criterion for him was that the machine needed to be flexible. As we have seen, the existing machines failed because they could not handle bottle changes very well. By rearranging the photo cells, new bottle types could be added with relative ease to the machine's list of acceptable containers. While this modification had to be implemented by a Tomra service person, it was a big improvement over the competing models with fixed acceptable bottle types. Tomra's prototype could easily handle the new one-liter bottle, which Tore highlighted in the patent application as something that distinguished their machine from the competitors.[30]

PREVIOUS PATENTS

Tore Planke had defined the engineering problem to be solved. But why did he choose to use such advanced technology for a task that so far had been deemed appropriate for "scruffy electronics?"[31] Like many inventors before him he started his research by looking at previous related patents.[32] Tore knew how the Wicanders machine and the Tveitan machine were constructed—the empty-bottle committee had provided him with technical drawings of these machines. His investigation revealed what the empty-bottle committee had already stated: current solutions were inadequate. Tore concluded that none of the existing patents "showed a navigable way" of handling of empty bottles in a more rational manner.[33] The current machine designs could serve only as a template on the most basic level, as a machine that received a bottle and gave back the deposit.

U.S. inventors had attempted to solve the grocers' bottle problem for a long time. Elmer M. Jones filed a bottle return machine patent in 1920.[34] Jones was an inventor who filed several patents on vending machines and train signaling devices, among others. Most of his devices somehow involved signals and feedback. Samuel J. Gurewitz of New York filed patents for bottle return and handling machines in 1952 and 1953.[35] Bruce Garrard of Atlanta filed a patent for a bottle vending machine in 1954.[36] These machines were all mechanical and gave the customer either a direct cash payment or a deposit receipt. Garrard's machine also sold full soda bottles; the empty bottles were stored in the slots that opened when full bottles were sold. This allowed for a compact machine—unlike Gurewitz's large and complex design. All the patents explained the need for bottle return machines in economical and infrastructural terms. Gurewitz added a comment on the problems with manual bottle handling: "The customer is required to go through a bothersome transaction for a few cents, the pleasantness of which may depend on the mood of the clerk."[37] After this spurt of bottle return machines in the 1950s, few machines of this kind were patented until the late 1970s in the United States.

The patent search made it clear to Tore that nobody had addressed the problem in what he called a "serious high-tech way."[38] Existing solutions always implemented low-tech, mechanical solutions with very limited performance. By contrast, he decided to design his machine with the "best available technology," which for him seemed the logical thing to do, considering his background in cybernetics and mechanical engineering.

TORE PLANKE'S PROTOTYPE RVM

Tore's preliminary experiments showed him that his idea was viable. After drafting a design for the first machine, he calculated an estimated manufacturing cost which the brothers presented to the empty-bottle committee. The committee was certainly pleased with the proposed RVM, concluding that it

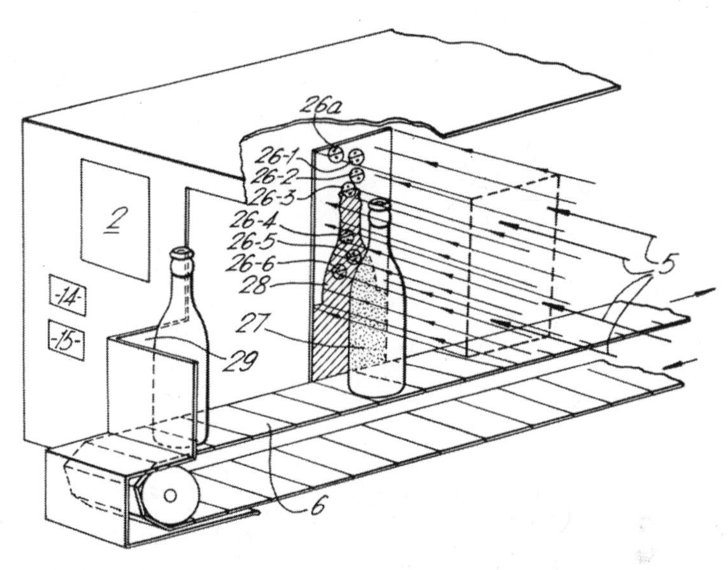

Figure 3. Contour recognition illustration from Tomra I patent application. Tore Planke, "Apparat for automatisk mønstergjenkjenning og registering av tomflasker," Norwegian patent no. 126900, filed December 14, 1971 and issued April 9, 1973.

"seems to satisfy the most urgent technical demands concerning handling of empties."[39] They furthermore agreed to subsidize the prototype by paying for the materials, which amounted to about twenty thousand kroner.[40] The brothers did all the assembly work free of charge. Petter hired a workshop to make the metal frame for the machine, while Tore provided the other components.

Tore filed a patent for his machine on December 14, 1971 (see Figure 3), and the brothers installed the prototype in Aage Fremstad's store the night of January 1, 1972, as described at the beginning of this chapter. During its first seventeen days of operation, customers returned thirty-two thousand bottles without problems.[41] This averages to about two thousand bottles a day, which is a significant amount.[42]

The Planke brothers invited the empty-bottle committee to visit Fremstad's store and examine the machine after a meeting in January. What they saw must have satisfied them. The final report from the technical committee praised the machine for being "reliable, easy to maintain, and meets the safety demands of both retailers and consumers."[43] A summary from the Oslo Grocers' Association enthusiastically stated that "the machine registers bottles so precisely that it will not give a refund for 'unknown' bottles."[44]

The committee considered its problem solved, assuming that the Planke brothers would produce and sell this machine at the estimated price of 23,500 kroner (excluding taxes).

JUDGING THE MARKET

Petter and Tore Planke had succeeded in developing the prototype RVM that the empty-bottle committee wanted, but they were still far from establishing themselves as a company. Aage Fremstad's store soon became a showroom for the Planke brothers. Petter Planke's long experience as a salesman had taught him that positive statements from grocers did not automatically translate into sales. After all, the Plankes' machine would be about twice as expensive as the currently available solutions. In addition, the Planke brothers did not have the capital to start producing more RVMs. They needed actual sales before they could take the chance on starting a company to produce the RVM. Petter Planke selected a group of grocers for a demonstration in Fremstad's store. These were all grocers he knew and whom he thought were most likely to purchase a RVM. He must have made good choices in selecting his audience, considering that seventeen of the grocers agreed to buy a RVM from him.[45] Installing the machine in Fremstad's store turned out to be a smart business decision.

The Planke brothers did not decide to start a company to produce the RVM without closely examining the market for it. The grocers' need was evident and the Planke brothers were convinced that their RVM could cover the grocers' need; the demonstration and the subsequent orders proved this. However, the Planke brothers looked further than just the backrooms of grocery stores when judging the viability of their product. As shown by international developments, especially in the United States, reusable bottles seemed to be giving way to disposable bottles. One result of this development could have been that grocers would not be required to collect them any more. Petter Planke called this realization the first crisis in Tomra's history. They knew they had a good product, but societal trends threatened to make the RVM obsolete even before they had a chance to start their company. At the same time, the Planke brothers realized that more disposables would lead to an even greater littering problem. In the beginning of the 1970s, Scandinavian newspapers were full of articles and letters complaining about littering, often related to disposable containers.[46] As we saw in preceding chapter, the Norwegian Storting would pass several environmental laws that addressed waste and littering in the early 1970s. The Planke brothers decided that increasing environmental concern would lead to a larger market for the RVM in the future. Because of this, the fate of the RVM was from the beginning tied to the parallel development of consumerism and environmentalism.

Starting an Entrepreneurial Company

The Planke brothers took the plunge and started a company. On January 17, 1972, Petter quit his Antonson Avery job and founded a company to produce and sell the Tomra I Reverse Vending Machine, short for TOMflaske ReturAutomat (Empty Bottle Return Automat). George Eastman's Kodak was a direct influence on this name.[47] The Planke brothers wanted a name for their machine that was short and sounded good, yet did not mean anything, just like Kodak. It also had to be easy to pronounce in any language without picking up any negative connotations on the way.[48] While they initially named the company Petter Planke A/S, they changed it to Tomra Systems A/S in 1973. Until the spring of 1973, Tore remained in his job for Norcontrol—another company partly springing out of the NTH cybernetics group—designing automated ship navigation systems during the daytime; at night and on the weekends, he worked on the RVM prototype.[49]

In the beginning, Tomra was a family-run business. Tore and Petter provided nine thousand kroner and their parents chipped in the last one thousand kroner necessary to start a joint-stock company. In April 1972, they rented an old World War II German barracks in Asker. The entire Planke family helped get the premises ready over the Easter holidays to produce the first Tomra RVMs. Petter was the managing director and salesman for the company; Tore was the technical director; Petter's wife, Grete, was the office manager; and their mother, Vivien, was their switchboard operator and warehouse manager, as well as the production assistant, soldering the occasional wire. Their father, Sverre, helped with some of the foreign correspondence.[50] In addition to the family, the Plankes hired two employees in 1972.[51]

In the first year of operation, Tomra sold twenty-nine RVMs in the Oslo region. These machines received about three million bottles during that year.[52] This is an average per machine of about one hundred thousand per year or 283 per day.[53] Tomra made upgrades to these machines at no extra cost to the grocers to increase functionality and reliability. Tomra was highly dependent on goodwill and the reputation that was built among the grocers to secure further sales. The Planke brothers knew that creating a strong network of RVM proponents would be key to the success of the technology and their company.

THE PLANKE BROTHERS presented their machine to potential customers through demonstrations, word of mouth, and direct sales. In 1973, Tomra made their first advertising brochure. The brochure had a picture of the RVM, highlighting a detail from the printer that was part of the machine. The text reads, "Avoid losses—Save time. Empty bottle machine that gives you a printed receipt with calculated deposits." The last line states, "Tomra—An electronically controlled empty-bottle machine," emphasizing the difference between the competitors' old mechanical machines, and Tomra's new electronic model.

During the following years, Tomra introduced a new element to their advertisements: satisfied customers. The front page of their eight-page, full-color brochure *Automatic Bottle Collector—Protects Environment and Profits* from 1974 shows a picture of a smiling woman holding a refund receipt. The accompanying text stated that "printed receipts showing deposit refunds give reliable control—and satisfied customers." By reiterating the fact that Tomra's machines gave printed receipts, the Planke brothers wanted to further distinguish their machine from Tveitan's model.

The brochure also focused on reliability, control, and accountability. It contained two statements from grocers who had installed the machine. One manager explained how installing the RVM had revealed that there had previously been a 30 percent loss on bottle returns. Through reducing the losses, the cost of purchasing a RVM was recouped in less than seven months. Another store manager said that the RVM had helped reduce his store's bottle return staff from six to two people in the high season, when they would receive up to eight thousand bottles a day, and that "there is no doubt that Tomra is earning money for us."[54] In this way, the grocers acted as witnesses to the RVM's efficacy. The RVM appeared as a "miracle machine" for the grocers. Tomra clearly framed the RVM as an economic and infrastructural tool to make the grocers' part in the recycling system as convenient as possible.

Tomra designed its bottle return machine to appeal to grocers by giving it a functional, if not exactly visually appealing, design. The company relied on economical and rhetorical arguments to convince grocers to invest in RVMs. By highlighting how the grocers could delegate the task of bottle handling and counting to the machine, Tomra addressed the grocers' dislike of this task. This strategy extended to include store customers. Printed user guides displayed in the stores helped customers use the machines. Tore Planke attempted to minimize user interaction with the technology by making it as simple as possible to use, "so that there was no resistance."[55] In this way, convenience served as a powerful tool for influencing user behavior and increasing demand for the machine.

Increased publicity made sales a lot easier in the following years. Tomra installed a total of seventy machines in 1973 and eighty in 1974, which increased the turnover from 800,000 kroner in 1972 to 2,100,000 in 1973 and 4,800,000 in 1974. This shows that Tomra's first two years of existence went reasonably well. Still, it sold its machines one by one, and much hard work lay behind every sale. Tomra presented its machines in several trade fairs in the United States, Norway, and Germany. In 1973, about 25 percent of its income came from sales of machines to the United States, Holland, Germany, France, and Sweden. In 1974, Tomra RVMs were in use in ten countries.[56] This shows that although the Planke brothers had designed the machine specifically for the Norwegian market, grocers in other areas faced the same challenge of handling empty bottles. As we will see in next chapter, growing international demand would require some technological adjustments.

Solving Infrastructural and Logistical Problems

The infrastructural problem of handling bottles in grocery stores formed the immediate background for the Tomra RVM. The machine was intended to take the place of human labor for handling and counting bottles. Petter and Tore Planke made these problems their starting point when deciding to design and sell a high-tech RVM to compete with the existing mechanical solutions. The combination of Petter's marketing experience from years of working as a salesman in the retailing industry and Tore's technological expertise in cybernetics and mechanical engineering proved fruitful. They were able to convert the need into a defined engineering solution that then could be marketed and sold as a commercial product.

Although the economic and logistic concerns of grocers dominated the machine's development and distribution, the Plankes placed the machine in a broader cultural context from the beginning. They recognized an international trend of environmentalism, stating that "all over the world, development heads towards increased protection of natural resources. Higher deposit values and restrictions on disposable containers clearly show this. Our machine has therefore been received with great interest in many countries." [57] An increasing visibility of litter in cities, along roads, and in nature worried many consumers. The freedom and convenience promised by consumer society were giving hints at its dark side. The Planke brothers hoped their technology could provide a solution to this problem as well. However, Petter Planke recognized that arguments about nature and protecting the natural environment would not help them sell to grocers.[58] Such concerns could not sell machines to these individuals, who focused on more immediate problems.

As a consequence, the RVM was designed, marketed, and sold as an infrastructure machine through the use of predominantly economic arguments, which identified the RVM as a critical node in the recycling system. It would still take many years before the RVM could stand on its own as an environmental machine. Yet the Planke brothers' machine found its niche in the Norwegian beverage container system by improving the infrastructure. When they attempted to sell the machine in international markets, they faced a whole new world of bottles.

Chapter 4 A World of Bottles

In the spring of 1978, Tore Planke and Georg Ås, Tomra's international-service engineer, installed a prototype of their new high-tech self-learning reverse vending machine, called the SP, in a small grocery store in the countryside in central France. It was important for them to test this machine in France, since this market had an enormous number of different bottle types. The diversity and complexity of French bottles posed the ultimate test for Tomra's next-generation RVM, a machine on which the company hedged its entire future. On the other hand, they also wanted to test it in a low-profile location where few could see the machine, test it out, and figure out how it worked. They installed the machine and the accompanying software and trained a local service technician. A month passed by—the machine worked with just a few small glitches. Then, in May, Tore received a frantic phone call from the service technician: "Tore, c'est pressant! Come to Paris immediately! We have a big problem with the SP machine!" Tore was surprised by this, as the machine was supposed to be in a remote corner of the countryside. It turned out that the service technician had taken the machine out of the store and displayed it at the largest supermarket exhibition in France without informing Tomra. Grocers attending the exhibition immediately saw the potential of the new machine and wanted to purchase one right away. As a result, the sales of Tomra's current model, the Tomra I, dropped to almost nothing in France, their most important export market. While the self-programmable machine was yet not ready for production, hardly any French grocers wanted to purchase a machine based on old technology after they saw the new prototype.[1]

This story hints at how important international markets would become for Tomra by 1978. Tomra had success selling their first machine in Norway, but as their market expanded internationally, they had to rethink their vision for Tomra and its machines. New, more flexible technologies would become the key to Tomra's success internationally, but as Tomra's experience with the rogue French technician indicates, it was an uphill struggle to get there.

With the Tomra I, the Planke brothers had addressed the grocers' direct bottle problem. But the Tomra I was a machine custom built for the standardized social democratic Norwegian bottle; it struggled when facing consumer diversity and international markets. This chapter will examine two problems with Tomra's first RVM solution and the company's attempts to solve them in the late 1970s and early 1980s. First, it was hard for Tomra to compete in Norway because of the high cost of the company's machines. Tore Planke had opted for what he considered the best available technology for the Tomra RVM, but this made the machine's cost prohibitive for many small grocers. So Tomra developed a second, "light" machine with older, less advanced technology in order to lower the cost for the domestic market. Second, their machines were noticed internationally, prompting grocers in other countries to install Tomra RVMs. France quickly became their largest export market, but it also posed a challenge because of the large number of bottle types in circulation. Moving into new settings increased by orders of magnitude the original bottle problem that Tomra's RVM had been built to solve. So Tomra turned to even higher technology to meet the demands of international beverage container recycling systems.

The two problems of cost and internationalization highlight the challenges posed by the materiality of bottles. The variety in shape, weight, opacity, and material all determined the ability of the RVM to identify and handle the bottles. In the process of meeting these demands, Tomra not only refashioned its technology, but also created a new vision for the RVM and its place in recycling systems in domestic and international markets. The technological choices taken by Tomra and the Planke brothers set them on a course that would lead them out of the backrooms of grocery stores and into the design and implementation of large-scale international recycling systems centered on the RVM.

Facing the Competition at Home

The Planke brothers' efforts in targeting the Norwegian grocers and their customers paid off and kept the company rapidly growing the first few years. The new higher deposit values following the 1974 bottle deposit regulations increased the number of potential customers, as did the ever-increasing beverage consumption rates. An increased number of bottles flowing through grocery stores meant that the need for bottle-handling solutions grew. However, Tomra faced several challenges in reaching a larger domestic customer base. Price was the first and most immediate of these.

While the first Tomra RVM was considerably more advanced than its competition, it was also very expensive compared with its main Norwegian competitor—Arthur Tveitan's bottle return machine. Tomra's original price estimate for the Tomra I RVM was 23,500 kroner (approximately $3,000).[2] Tveitan's much simpler machine cost about half as much.[3] Since Tomra could not compete on price, it had decided to focus on features and reliability in their marketing of the machine, capitalizing on the high-technology approach of Tore Planke.

These benefits more or less evaporated when Tveitan launched a new model that had some of the same features as the Tomra machine. Tveitan felt the need to upgrade only after their customers began switching to Tomra's more advanced machines. Tveitan's new RVM, the ATAS, could be reprogrammed to use different deposit values, and it also had a control mechanism where the grocer could get a transaction summary. While Tomra's bottle recognition mechanism was still more advanced, Tveitan had copied two of the key technologies used in the Tomra I and adjusted his selling points to closely resemble those used in Tomra's brochures. Both companies presented their RVMs as a technology for solving grocers' infrastructural problem. In Tveitan's six-page advertising brochure for the ATAS bottle return machine, the front page featured a photo of the machine, showing both the front-end bottle collection and registration mechanism—a cash register on top of a metal box with a slot for bottles—and the backroom collection table. Ease of installation was one of the most important arguments in the brochure, which included technical drawings of the alterations that needed to be made to the store wall and backroom. The brochure described the machine in technical terms, explaining how it registered bottles. Using a standard cash register as the main interface to the machine probably served to make it familiar to grocers. The brochure even emphasized that "if the deposit system for some reason should disappear, the registration device can be disassembled and used as a regular cash register."[4] With this statement, Tveitan clearly appealed to the frugality of the grocers. Even though Tveitan's machine was less sophisticated than the Tomra machine, it did the job well enough and kept Tomra out of Tveitan's established user base, especially in the Co-op stores. For these customers, price was as important as features and performance.

In 1973, Tomra decided to face the competition from Tveitan by launching the Tomra Junior. The Junior was a low-price machine based on mechanical bottle recognition—just like the competing technologies that Tore Planke two years earlier had deemed insufficient—instead of the advanced optical recognition that distinguished the Tomra I. The company sold the Tomra I by its being more advanced and more reliable than the competing machines, whereas the Junior imitated the competition. It was designed to fit in smaller stores, as it was installed as a free-standing unit with a small footprint and a removable collection table, which meant that it did not require installation in the backroom. Unlike the competing machines, however, the Junior measured the diameter of the bottle, and not the height. Tore Planke discovered that this was a more reliable way to identify the bottles on the Norwegian market. The Junior also used cheaper and less durable materials. Instead of the solid metal front that the Tomra I had, the Junior had a laminated wood front on top of a steel frame cabinet. Inside, the Junior featured the same printer and the same computer board as the Tomra I, allowing Tomra to reuse some of the technological solutions they already had developed. All these cost-cutting choices allowed Tomra to sell the Junior for eight thousand kroner less than the more advanced Tomra I.

At first, the low-budget machine was a big success. Tomra sold fifty machines just at the trade fair where the Junior was first introduced. As a result, the Junior opened up a new market for Tomra, yet the Planke brothers were ambivalent about using such a low-tech approach. In 1974, only one year after the introduction of the Junior, Tomra stopped its sales because they did not think it met "future demands." Tomra accepted the loss of some sales in Norway "to a competitor with an affordable machine with mechanical registration,"[5] most likely referring to Tveitan. This choice demonstrates that Tomra thought of itself as a high-tech company and the bottle problem as something that could be properly solved only by advanced technology. In Tomra's view, the RVM needed to do more than simply replace some of the manual labor in the backroom.

Yet Tveitan continued to be a serious competitor, leading Tomra to reconsider the Junior. In 1975, Tomra reintroduced the Tomra Junior, but with one key technological improvement: a new printer for the customer receipt. The old printer had been purchased from Switzerland, and had fifty to sixty movable parts. After two to three years in operation, this printer represented 50 percent of all service costs for the machine because of sticky soda and beer residues. Tore Planke developed a new thermal printer with only two movable parts, which reduced the production and maintenance costs significantly. They also made some changes in the mechanical construction that made the machine very sturdy—to the degree that the Junior hardly required any maintenance at all.[6] A reliable low-cost machine thus became more feasible.

By focusing on long-term maintenance and service costs, Tomra managed to make the redesigned Junior a successful product. It became popular in small stores and workplace cantinas, as well as in the cost-conscious stores that Tveitan had as his main customers. By adopting "old" technology—and doing it well—Tomra managed to stave off most of the competition. While Tomra wanted to sell high-tech solutions to Norwegian grocers, many smaller grocers just did not need—or were not willing to pay for—such advanced technology. Tomra made the Junior to resolve this gap between the ambitious high-tech ideals of the Planke brothers and the everyday situation of these grocers. In other words, while Tveitan upgraded his machine, Tomra downgraded in order to find the optimal price/feature combination for Norwegian grocers.

A World of Bottles

The second challenge Tomra faced in the early 1970s required an opposite strategy. As a result of its success in Norway, representatives for international distributors who wanted to sell the RVM in different markets soon contacted Tomra. The Planke brothers had clear economic incentives for internationalization. They had chosen an area with few established competitors and thus found a niche for themselves, but the narrower the niche, the smaller the home market is. In order to keep growing and to find new customers for their machines, Tomra needed to look outside the limited Norwegian market. However, bottle problems in other countries were in many cases much more complicated than the Norwegian problem. This encouraged Tomra to develop and introduce more advanced and more flexible technological solutions in their RVMs.

Sweden, which had a deposit system similar to Norway's and over seven hundred million containers produced each year, early on became Tomra's most reliable international market. As did Norway's, the Swedish government worried about the littering problems that disposable containers might create.[7] A 1969 government white paper titled "A Cleaner Society" blamed the dramatic rise in littering on disposable packaging, though changing leisure habits and increased mobility had to take some of the blame.[8] Sweden instituted a tax on both returnable and disposable beverage containers in 1973, leading to higher prices on disposables and a drastic increase in the deposit on returnables from ten to forty öre.[9] The old deposit had been lower than the fifteen öre price for new bottles, which led to a steady decline in the use of refillable bottles.[10]

The new tax strongly favored reusable containers; since the tax was only applied to the container once, reusable containers were cheaper in the long run. While some alternative disposable containers like the cardboard/PVC

Rigello bottle, single-use glass bottles, and steel cans had gained a small market share, the refillable bottles came back as the dominating container on the Swedish beverage market.[11] As we'll see in the next chapter, disposable containers did not really gain momentum until the early 1980s. Within just two years of the increased deposit value, the use of returnable bottles increased from 79 to 91 percent.[12]

To help Swedish grocers with their bottle problem the Swedish beverage container manufacturer Plåtmanufaktur AB (PLM) contacted Tomra in 1973 with an offer to sell its bottle machine in Sweden. Through this move, PLM indicated that it wanted to be involved in the entire packaging value chain, though the strong public backlash against beverage container littering may have influenced its decision to expand into the collection business. In the first year, PLM sold twenty machines for Tomra. While it was not a huge number, the Planke brothers found this arrangement very satisfying.[13]

Tomra's international breakthrough order came when Systembolaget (the Swedish Wine Monopoly) ordered one hundred machines in 1974, at a time when Tomra typically sold its RVM one by one. Systembolaget already had a semiautomated solution with a conveyor belt installed in all three hundred of its stores, so it wanted Tomra to custom make a RVM that would use this conveyor belt instead of the regular transporter. It took Tore Planke less than two months to adapt the Tomra I to these specifications.[14] Just as in Norway, Tomra worked closely with its customers to create solutions to their specific problems. This was also the company's first experience with connecting its RVMs to existing technological systems, albeit a small-scale one. Winning this large order was Tomra's real windfall and gave the company some financial freedom. Tomra outgrew the old barracks where they started, so the Planke brothers started building a new, expanded factory in Asker early in 1975.

Customization of the machines worked well in the case of Systembolaget, but proved untenable in many European markets. By 1976, Tomra exported 60 percent of the RVMs they built.[15] As Tomra became more international, reprogramming the Tomra I for individual stores became increasingly problematic and time consuming. As early as 1972, Tomra recognized that "the registration programs for other markets became very demanding because of the large variety of bottles."[16] For Tomra to continue growing as an international business, they needed to refine their technology.

Teaching the Machine to See

The French market was the first that severely challenged the technical capabilities of the first-generation RVM. Tomra had originally entered France in

1973 when the French subsidiary of Mead Packaging, a large American packaging manufacturer, contacted Tomra with an offer to sell the RVM there. Tomra eagerly accepted the offer. There was just one problem. The Tomra I could be programmed to accept eight different bottle types through a single opening, which worked well in Norway where there were just a few, standardized bottle types. In France, however, Tomra faced a wholly different situation.

While three-quarters of the beer and wine bottles on the French market were reusable, there was no such thing as a standardized French bottle. On the contrary, France had more than three hundred different bottle types and about one hundred of these carried deposits. French law had mandated deposits on refillable containers since 1939, but there had been no efforts to establish a cooperative system with standardized bottles, as in Norway. One study concluded that "due to the wide variety of wines and spirits it is difficult to establish an economically viable system to return the bottles to the original bottler."[17] The old reuse packaging systems degraded rapidly in France as bottlers and brewers considered one-way packaging as a way to avoid the mandatory deposit on refillable containers. First, the reusable glass bottles in which water was sold were replaced by PVC bottles and one-way glass in the 1970s. Second, the beer bottle system declined in the mid-1970s, as one-way glass began to take over.[18] So when Tomra tried to enter the French market in the early 1970s, the entire bottle situation was somewhat in flux, with large quantities of different new and old bottle types in use.

France's messy bottle situation was no less inconvenient for the consumers than for the grocers. French stores would accept only the bottle types that they themselves sold. In an interview, Petter Planke told the story of how Georg Ås, Tomra's engineer supporting the French market, found that the French customers claimed the machines were stealing bottles. After installing a Tomra I in France he observed a customer insert twelve bottles. When the machine paid for only four of them, the customer said, "This machine is stealing bottles." The store owner replied that only four of them were refundable there—"Look at the sign over the machine there. You paid your deposit in another store." Then the customer showed another crate to the supermarket owner, with twelve more bottles. "How many of these will I get a deposit for?" he asked. "Three bottles," replied the grocer. "Merde!" shouted the customer, who then dumped out the whole crate, crushed all the bottles on the floor, and left the store.[19] As the front end to the entire bottle recycling system, the RVM came to take the blame from many frustrated and confused consumers.

Because of this sticky situation, every RVM had to be custom built for the individual grocery store in order to program exactly which bottles the machine would pay refunds for. Tomra's service personnel thus had to travel all over France to install the machines. Through their first-generation RVMs, Tomra had identified a large potential market in Norway and abroad, but it had come to realize that it did not have viable technological solutions for the export markets yet. To take on this challenge, Tore Planke had to rethink the technology behind Tomra's bottle machine.

Moving the RVM out of the Norwegian context in which it originated required switching to an entirely different technological basis for recognizing bottles. In a world of hundreds of different bottle types, the RVM could no longer rely on Tomra's engineers painstakingly and manually configuring it for all the different bottles. The machine needed to be able to operate and to learn independently.

A detailed look at the Tomra I bottle identification mechanism demonstrates why this machine depended so utterly on engineers. The Tomra I used halogen and photocell recognition technology, and could recognize up to eight different bottle types.[20] Tomra produced approximately twenty-two hundred such machines between 1972 and 1979. A technical diagram of the Tomra I shows the approximately seventy-five sensor units that registered the outline of the bottle. As the bottle moves through the machine, it is exposed to light rays passing through the bottle. The glass on the edges of the bottle will break the light, so that it doesn't hit the photocell sensor on the other side. As a result, some sensors will light up, while others won't. This is how the machine produces an outline of the bottle. The sensors marked with darker circles are the ones that are used for identifying the particular bottle drawn on the diagram. The documentation required for identifying all the different sensors and their placement composed a massive tome. Positioning the sensors in the correct place for recognizing bottles was a painstaking process that could not be performed by untrained personnel. Tomra's engineer had to physically place the sensors at the exact places that were unique for the bottle. This sometimes required fine-tuning the position down to a fraction of a millimeter, and would take up to thirty hours per machine.[21]

The bottle contour then had to be translated into a form that the machine could understand and act upon. Teaching the machine to recognize bottles in this way was a manual process, requiring a good eye for shapes and outlines. The same thing had to be done when reprogramming the machine. As such, it was a skill that was embodied in the engineers themselves and that could not be easily transmitted to the machine. For example, Tomra's engineers had had to travel around Norway in order to reprogram all two hundred machines the company had installed when the deposit rates changed in 1974. While the Tomra I was technically advanced, it was a form of high technology

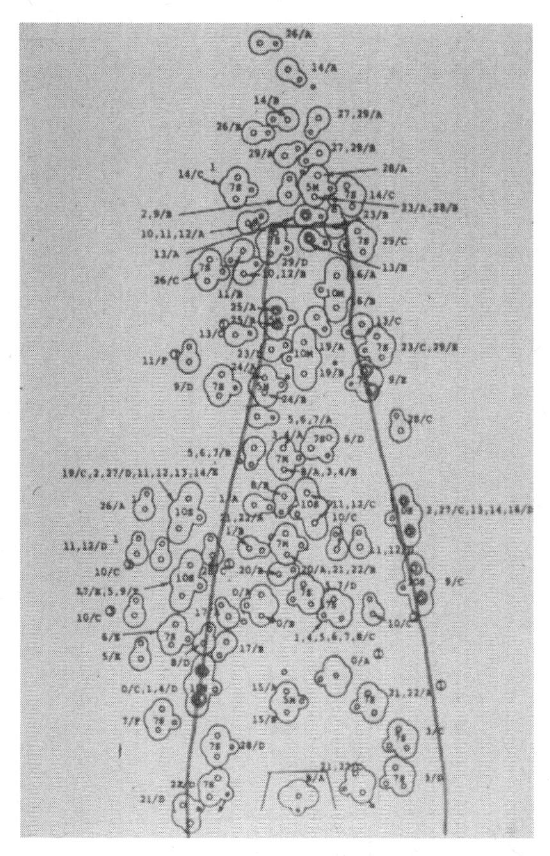

Figure 4. Technical diagram demonstrating the photocell bottle recognition of the Tomra I Reverse Vending Machine. The outline of a bottle is drawn on top of the photocell detectors. Courtesy of Tomra Systems ASA.

that required heavy maintenance by Tomra's engineers. Or to put it another way, there was much that was not automatic about Tomra's automats.

The limits of the technology used in the Tomra I RVM were beginning to hinder Tomra's further growth. At the beginning of 1975, Tomra therefore decided to develop a new generation of RVMs. The goal was ambitious: to create a machine that could be used in all markets and in all recycling systems and where the adaptation to the market's selection of bottles could be done by a self-programmable machine. This would be a more flexible and independent RVM that could operate freely in a world of bottles.

The conceptual development of the new machine was intense and focused. Tomra's engineers Georg Ås and Magne Storli spent a week with

Figure 5. Tomra engineer Magne Storli comparing bottle shapes. Programming the early RVMs to recognize refundable bottles was a highly manual process. Courtesy of Tore Planke.

Tore Planke working out the specifications for the self-programmable RVM in 1975. This small design team worked intensively during this week from 8:00 in the morning until 10:00 at night in a small conference room at a mountain hotel in the resort area Norefjell. The only interruption was the daily two-hour ski trip to a snow shelter they built, where they had lunch and one beer each, before they went back to the hotel, all while constantly discussing the RVM. When they descended from the mountain after a week of highly focused work with no telephones and no interruptions, they had the specifications for the self-programmable ready. Tomra had chosen the high-end technological path they would follow in the years to come—and none had walked it before them.

Their new machine, the Tomra SP (short for *self-programmable*), used microprocessors, fiber optics, and lasers—all of which were new untried technologies in 1975. Line scanners, which the old machine used, had proved to be too hard to reprogram. Small helium lasers, intended for bar code scanning in cash registers, had by 1975 become available for four thousand to five thousand kroner each. While this was very expensive—about the entire production cost of a Tomra IIIC, the last of the first-generation RVMs—Tore Planke connected the idea of using lasers with something he had read about the first CD players, which had recently been tested in labs in the United States. He realized that if it was possible to pack computer information on a small disk, it would be a huge commercial item—which meant that the laser would drop in price. Therefore Tore thought that the cost savings would be realized later.[22]

The engineering team faced two serious technical problems with the stability of the lasers used in the SP. First, most lasers did not have sufficient pointing stability, which meant that the beams would jump in different angles and not hit the detector. Since nobody had used lasers in this way before, the Planke brothers did not know whether they would succeed in working out the kinks—"whether the light at the end of the tunnel was a meeting train or the end," as Tore Planke put it.[23] If they could not get the SP to work, Tomra would not survive. By switching to so-called hard-sealed lasers, the engineering team finally solved this problem.

The second problem was how to detect the laser beams properly. The team decided that fiber optics sensors would work. Nobody had used fiber optics to detect lasers before. Finally, they decided to use a microprocessor to control the interaction between the lasers and the fiber optics sensors. The Intel 4040 microprocessor had just become commercially available in 1974. The Norwegian Institute of Technology had a lab that worked with microcomputers, and Tore Planke tapped into this knowledge. In the Tomra SP design process, he actively leveraged his connections to the Norwegian university–based R&D labs.

So when Tomra decided in 1975 to make this machine, it took some big chances: it used prohibitively expensive lasers; it used fiber optics, which was quite new and untried in this context; and it used microprocessors, which hardly anyone knew how to program yet. The lasers eventually got cheaper, even though it took longer than Tore had hoped. For several years Tomra was the largest European purchaser of Japanese lasers.[24] These decisions aligned well with the Planke brothers' wish to create what they called the best—not the cheapest—solution using cutting-edge technology. It also reflects their enthusiasm about the Tomra technology and their self-image as leaders of a high-tech company. The Planke brothers hedged Tomra's future on several untried technologies. A more established firm—one that did not operate in an

Figure 6. "The RVM that teaches itself which bottles you have to pay for . . ." This Tomra SP advertisement from the late 1970s emphasized the new machine vision by juxtaposing a bottle contour with an eye. The text clearly targeted grocers, who had to pay out money for returned bottles. Courtesy of Tomra Systems ASA.

entrepreneurial mode—might not have taken such risks when developing a new product. Without the Planke brothers' enthusiastic, aggressive, and sometimes naive attitude, Tomra probably would not have succeeded. It certainly would not have been able to expand internationally at this stage.

The Tomra SP worked by computer-based image processing and pattern recognition. The RVM did not look at the whole bottle and recognize it. Instead, a laser swept across the bottle as it passed through the machine. Two hundred and fifty-six sensors registered the bottle outline, generating twenty-five thousand identifying points, whereas the old machine only could make one hundred. The computer translated the contour into a series of numbers unique for this bottle, and compared the sequence to a database of known

Figure 7. Tomra's "bottle-o-teque" contained more than two thousand different bottles from all over the world. It was a clear manifestation of Tomra's international expansion and the challenges the machine faced when moving into new markets. Courtesy of Tore Planke.

bottles in order to decide whether the bottle had a deposit.[25] Bottles always deviated somewhat from the norm, so the SP had a certain tolerance for variations built in. This database was an important part of Tomra's software and was generated and maintained through Tomra's enormous bottle collection, containing thousands of bottles from all over the world. This collection eventually grew so large that the beverage industry sent representatives to Tomra when they planned new bottle types because here they could find exemplars of most bottle types on the market.

Tomra designed the SP to be adaptive and self-learning. When a grocer needed to add new bottles, he or she could insert a bottle through the hole in the wall and the computer would scan it, calculate a new identification number, and add it to the database. Just like the Tomra I, the SP would register the total number of bottles returned, as well as the total sum refunded that day. It also included monitoring and self-diagnosing in case of errors and mechanical problems. Furthermore, the SP had a button that the consumer could press to call for assistance, an optional backroom solution, an optional keyboard for registering deposit values, and an optional video camera by which the store clerk could monitor the bottle returns. Now Tomra could produce one model for all markets and customize it through the software, eliminating the need for time-consuming and tedious adjustments by engineers. The availability of optional accessories also enabled grocers to further customize their installation to their individual needs.

With the SP model, Tomra responded to the material input of new bottle types by developing an even more technologically advanced machine. User requirements were fed back into the design process and led to new technological solutions. These solutions, however, were still rooted in the backrooms of grocery stores. As the recycling system became increasingly dependent on technological solutions, these technologies needed to address more and more user needs.

Financing High-Tech Choices

Tomra spent significant research and development money to create such advanced machines. They invested close to three million kroner in the development of the SP, a remarkable sum for a company with eight million kroner in annual sales. This expenditure had a negative influence on the company's annual accounts, leading the Planke brothers to explain these results in the 1977 annual report, which stated that the company had put much effort into "securing the future through product development and through changing from traditional electronics into microprocessor electronics" despite the difficulties of raising capital for investments. By highlighting how the SP technology was flexible and future-proof, the Planke brothers managed to preserve customer confidence.

However, they also managed to make their current technology obsolete. Tomra's high degree of R&D investments inspired newspapers to describe Tomra as a company with "a philosophy of making their own machines obsolete before their competitors can."[26] With the Tomra SP, this strategy worked too well. As early as 1975, the annual report announced that the "principles behind the machine were tested" and that a prototype would soon be installed.[27] The 1976 annual report stated that the machine now was ready for production and that sales would start in late 1977.[28] Newspaper articles also presented the new bottle machine long before it was ready for sale.[29] In effect, Tomra did its product development in public. In 1978, prompted by the French outing of the prototype SP, Tomra realized that the high expectations built by the company's publicity work caused sales of the old machines to drop—long before Tomra was ready to mass produce the SP.[30] This put it in a tricky situation and forced it to launch the SP before the machine was finalized.

In the years between 1975 and 1978, Tomra completed the switchover from traditional electronics to microcomputer electronics. Developing the microchip controller card—the brain of the Tomra SP—had been complicated and expensive, but after finishing this work, Tomra had the equipment and know-how to begin using the SP technology in all its machines. The previous three years had been hard on Tomra's finances and the company was balancing on a very thin line. The 1978 annual report emphasized that this development had to be seen as an investment, and not as a cost, promising better results for next year. And they delivered: Tomra produced almost six thousand SP units between 1978 and 1988. The SP was particularly successful in France, where sales doubled in 1980.[31] The self-programmable functionality and high flexibility was custom made for markets like France.

Tomra secured patents as part of their strategy to earn revenue from its machine technology. The Tomra SP was not only made to recognize bottles; it was also a high-tech business machine made to hinder competition. "We have a product that through a combination of mechanics, electronics, and software is hard to copy," said Tore Planke in 1979. "Patent applications often take a long time, and the product may be obsolete before the patent is granted. But we consider patents to be very important, both as a brake for possible competitors and as an important document when entering into or maintaining international agreements."[32] Tore Planke's first patent application for bottle reception was registered by the Norwegian patent office on December 14, 1971, and approved in 1973. Having this patent helped hinder a number of competitors, including Tveitan. In addition, Tore filed patent applications in twenty countries, demonstrating how the company already in the beginning looked beyond the Norwegian market.[33]

The threat of patent lawsuits was an effective tool for Tomra. In 1978, Tomra confronted Hugin Kassaregister AB, a Swedish company, with patent infringements. Hugin had taken Tveitan's bottle machine and adjusted it to the Swedish market. As part of this customization, they replaced the mechanical recognition mechanism that Tveitan used with a photocell technology very similar to Tomra's patented solution. Tore Planke met with Hugin's top management, documented Tomra's patent, and gave Hugin with choice of a patent lawsuit or a cooperative agreement. Hugin caved in, stopped selling and marketing Tveitan's machines, and began using Tomra machines instead. This gave Tomra a channel into all the supermarkets in northern Europe, because Hugin had a large distribution network in twelve countries. The Planke brothers effectively used the patent as a stick and the SP technology as a carrot in recruiting Hugin as a distributor.[34]

Refining the SP Technology

Tomra soon built the SP technology into other products. In 1980, Tomra started developing a SP bottle crate return machine in cooperation with Sintef, Tore Planke's former employer. Halvor Wergeland was Tomra's engineer in charge of the project. Tomra had previously produced a semiautomatic crate return unit—the Tomra CRM—featuring a video camera linked to a monitor at the checkout counter. The store staff had to manually check that the crate was full of empty bottles for the refund receipt to be printed. Although Tveitan had tried to update their machines with the ATAS, it decided to leave the RVM market after Tomra introduced the CRM.

The automated crate recognition system, or CRA, improved on the CRM by removing the need for manual approval. It used an ultrasound 3D scanner to achieve 95 percent accuracy when scanning crates of returned bottles for refund. The crates were scanned from the top, measuring the size of the crate as well as the size of the bottles it contained.[35] The CRA was programmable and self-learning, just like the SP. The Royal Norwegian Council for Scientific and Industrial Research (Norges Teknisk-Naturvitenskapelige Forskningsråd) innovation program awarded the project 480,000 kroner in support. Tomra's decision to create a crate machine with the same type of technology used in the SP reiterated the company's commitment to its high-tech machine vision.

Creating flexible machines that could fit into various system configurations continued to be a primary motivator behind Tomra's product development decisions. Although the SP was incredibly successful, Tomra quickly began working on an upgraded model in 1982.[36] The Tomra 300 was based on a modular design and used fewer components than the SP. In addition, the machine featured an improved bottle scanner using semiconductor lasers instead of the old helium-neon lasers. These lasers were the same ones used in CD players; Tore Planke's technological gamble on using lasers finally paid off. What had previously been prohibitively expensive could now be bought cheap and in large quantities. Furthermore, the machine was designed with a series of optional accessories, including one for recognizing crates. This allowed for even more customized RVM installations. The Tomra 300 was the first machine that could handle glass and plastic bottles as well as crates. By making these design choices Tomra wanted to increase the versatility and flexibility of the machine. Such design choices also tied into larger political controversies about beverage containers and market structures in the beverage industry, as we will see in the discussion of plastic bottles in Chapter 7.

Until the mid-1980s, Tomra's research and development processes centered on Tore Planke and his engineering team. Tomra was an engineer-managed company and operated in what we can call an entrepreneurial mode of product development. Economist Joseph A. Schumpeter defined entrepreneurs as exceptional individuals who could develop a new product, market, organization, or process, and then proceed to implement it in the economy.[37] The Tomra group had an extensive technological expertise in mechanical engineering, cybernetics, optics, and computer science, which they sought to translate into products that could solve problems related to the handling of empty beverage containers. Most of these solutions centered on Tomra's patented optical pattern recognition technology. The development of the Tomra 300 was originally intended to consolidate what they had learned during the production of the Tomra SP to make production easier and facilitate use of accessories.

The development of the Tomra 300 took five years and spanned a change in management (to which we will return in the next chapter). Tomra's annual reports are much less explicit about the development of the Tomra 300 than they had been when Tomra created the SP. The 1982 annual report promised an increase in the R&D budget for 1983: Tomra budgeted a total of seven million kroner for the development of new products. The Tomra 300 was not specifically mentioned, but most of the budget was probably intended for the development of this model. In 1984, Tomra established a research group to "catch and improve the different technological challenges that push from all directions" and to find "new methods" in two- and three-dimensional pattern

Figure 8. The Tomra 300. When Tomra introduced the Roy Tandberg–designed machine in 1987 after five years of development, it won the Mark for Good Design from the Norwegian Design Council. Courtesy of Norsk Designråd under Creative Commons Attribution–Share Alike 3.0 Norway license.

recognition.[38] The 1985 annual report merely stated that "we are nearing the completion of other projects that will have a considerable impact on Tomra's further sales development."[39] Tomra had obviously learned from its previous attempts at doing product development in public and did not want to repeat the mistake of making their current models obsolete.

At the same time, Tomra wanted to maintain its reputation as a constantly innovating company. Tore Planke wrote a short essay on technological progress in the 1983 annual report as part of this strategy. He presented Tomra's research group and stated that "we continually experience significant new break throughs" in pattern recognition and information technology.[40] Such rhetorical balancing acts were vital to Tomra's image as a high-tech company.

Tomra's public image also makes clear that Tomra focused on technological improvements as the way to new markets and better handling of the beverage container problem.

Tomra and the Norwegian High-Tech Industry

Tomra expanded rapidly through the 1970s, eventually taking a leading position in the Norwegian high-tech industry. It was not the largest company, but it was one of the fastest growing, most visible, and outspoken. The Planke brothers took a keen interest in a larger debate on the role the modern electronics industry could play in Norwegian society. As a result, investors, consumers, and politicians alike all began paying close attention to Tomra.

Through its rapid growth and the successful integration of the RVM in the grocery sector, Tomra clearly had an influence on ideas about recycling in Norway. For instance, the 1975 follow-up of the first governmental report on recycling and waste management (discussed in Chapter 2) presented RVMs as a method for making deposit systems technically and economically viable: "The handling of empty bottles and disbursement of deposits was previously labor intensive. A Norwegian company, Tomra Systems A/S, has for the last few years developed and marketed machines for this purpose, both in this country and abroad."[41] Such statements can be interpreted as a sign that Tomra's bottle machine was seen as a way of facilitating the practical implementation of environmental policy. A 1975 newspaper illustration indicated that consumers eagerly embraced the increased convenience the Tomra machines provided. Aftenposten's illustrator Sven Karsten Sønsteby made a drawing of three men running to the grocery store loaded with empty bottles. "Finally our grocer also got Tomra," exclaimed one of the men.

From the beginning of the 1980s, Tomra grew more confident as a company and the Planke brothers began to expand their influence to the wider Norwegian electronics industry. Tomra's high export rate earned the company the Norwegian Export Council's 1980 Export Product Prize for the Tomra SP—"the most advanced product [of its type] in the world."[42] The prize was a recognition of the strong international position Tomra had gained through exporting the SP. While Tomra had occasionally been featured in newspapers earlier, media attention exploded after Norway's Crown Prince Harald handed the company the prize. Articles and editorials in electronics trade journals indicate a certain pride that one of "their" companies finally was recognized.[43] The Planke brothers and Tomra could finally step out of the backrooms of grocery stores and into the public sphere. Most Norwegians had used their machines, but now they also knew who made them.

The Planke brothers actively used media to recruit interest in Tomra. Their participation in the public debate on Norwegian business and industry turned them into regular newspaper pundits. The brothers received two or three invitations a week from all kinds of organizations to give seminars and lectures, and they accepted as many of them as they could.[44] They became public figures who appeared on the covers of trade journals.

With his sales background, Petter Planke became a leader in the electronics industry organizations. The Electronic Industry Trade Association (Elektronikkindustriens Bransjeforening) invited him to be a board member from 1978 and then chairperson from 1981 to 1983—the first chairperson who was not an electronics engineer. He served on the executive board of the Norwegian Industrial Foundation from 1981, as councilor in the Norwegian Export Council from 1981, and as deputy board member in the Royal Norwegian Council for Scientific and Industrial Research from 1982. He actively used these positions to promote and advocate modern electronics. In one lecture, he asserted that "we must be increasingly conscious of what kind of society we want tomorrow and let us use modern electronics as a tool to create this society."[45] In interviews and editorials, Petter called for more higher education in electronics and argued that the adoption of new computer technology was not a threat to employment and for a coordinated Norwegian initiative for promoting entrepreneurship.[46]

Petter Planke admonished entrepreneurs to focus on specific user needs instead of just having good ideas.[47] Using Tomra as an example, he openly acknowledged that the company did not have the idea for the RVM at all; the grocers gave it to Tomra.[48] Petter insisted that technology was not created by entrepreneurs and inventors alone, but rather as "a process between market and technology. Our mission is to find solutions that cover needs by connecting the use of modern technology to the market."[49] Tomra's business philosophy was further refined as a result of the Planke brothers' engagement in the electronic industry. In 1983, Tomra used the phrase "market pull—technology push" about its approach to business and product development. However, to attain this mediating position between market and technology, Tomra had to be visible and reliable. This explains why Tomra worked so hard at both technology development and public image building; the two approaches were interconnected in a seamless web.

The Machine Vision

During the company's first ten years of existence, Tore Planke and the other Tomra engineers were concerned primarily with the insides of the machine—with making it solve grocers' bottle problem in a reliable fashion. In the early 1980s, they began to see beyond the backrooms of grocery stores and toward the intersection of technological and societal development. In Tomra's 1984 annual report, Tore Planke wrote a short text on society and technological development, arguing that "technology pushes on with an ever increasing strength—not just in electronics and computer science!"[50] The commentary discussed breakthroughs in material science opening up for new forms of beverage containers, something that could have an influence on Tomra's future. This tells us that Tomra actively kept an eye on international market developments and knew that its technology would have to respond to and even intervene in such larger trends.

In the process of teaching the RVM to see, Tomra also actively shaped the way the machine was seen by society and its users. Creating a self-learning RVM that could identify hundreds of different bottle types through advanced visual identification technology gave Tomra a strong international market position. This transformed the RVM from an advanced backroom technology to a truly high-tech infrastructure machine that had the potential to function in many different systems. The market success of the SP generation of RVMs signaled a new approach to Tomra's range of products, "the start of a whole group of products for recycling and compressing new forms of packaging."[51] The Planke brothers began to see themselves as system builders. "Tomra does not only sell RVMs, but complete return systems," asserted Tore Planke in 1984.[52] At the time this was somewhat of an exaggeration, but it was clear that the goal was to create larger systems for the handling of empty beverage containers. As the next chapter will show, developing recycling systems soon became an important strategy for Tomra.

Beginning in the 1980s, Tomra began framing its technology in a broader sense—as a technology for society at large, not just for the backrooms of grocery stores. Tomra's entry in larger discussions on the role of electronics in society also indicates how the RVM had matured as a technology and become integrated into Norwegian consumers' lives as a convenient way of handling bottles. The enlarged context in which Tomra situated its machine came to shape the company's efforts in influencing policy makers. The RVM thus became part of a new and larger network of actors and interests that stretched far beyond those of grocers and breweries.

Tomra's successes in the beginning of the 1980s had made the Planke brothers confident that they could grow rapidly to all of Europe and also to the United States—a market with virtually unlimited potential for Tomra, given the vast amount of beverages consumed every year. The next chapter will look at how this growth philosophy almost led to Tomra's demise.

Chapter 5 Can Cultures

On May 2, 1996, a *Seinfeld* episode called "The Bottle Deposit" aired on American television.[1] In this now-classic TV show, Kramer got one of his crazy ideas—to take empty cans from New York to Michigan so to claim the higher deposit. The New York deposit was 5 cents, while Michigan had a 10 cent deposit. The only problem with the Michigan bottle scam was that the transport costs were too high to make it profitable. Seinfeld's nemesis, Newman, who worked at the U.S. Postal Service, cracked the code. On Mother's Day, a total of five mail trucks drove to Michigan carrying mail. Four of them were full; the last one carried the spillover. Newman signed up for this truck so that they could fill the empty space with cans. Newman and Kramer then went about collecting empty bottles and cans, even to the point of stealing from homeless people. With the truck full of ten thousand deposit bottles and cans, they began the long drive to Michigan. In the usual twist of plots, their scam, of course, went awry (suffice to say that Seinfeld's stolen Saab and JFK's golf clubs were involved).

The *Seinfeld* episode shows how the awareness of bottle bills had entered mainstream American popular culture by the 1990s. What this story also tells us is that while some American states had bottle and can deposit systems, each state had different rules and different deposit values. This created the potential for scams like Kramer and Newman's. For instance, in 2006 two individuals and two corporations pleaded guilty to federal criminal charges of defrauding the Californian beverage container recycling program for millions of dollars. Between 1998 and 2000, they ran a complex scam of bringing cans and bottles from Mexico and nondeposit states and then claiming the Californian refund.[2] The lack of a national, coordinated deposit system that could distinguish between domestic and foreign containers made this possible.

From the story above, the differences between the complex American market and the smaller Scandinavian beverage container recycling context are obvious.

The bottle problem was international—meaning that there was a large potential market for reverse vending machines. While France and Sweden became important markets for Tomra early on, the Planke brothers really set their eyes on the United States—the largest beverage market in the world and thus a market of seemingly near infinite potential for their product, especially in light of the passage of state bottle bills in the 1970s. When Tomra attempted to sell their RVMs outside the Norwegian home market, they were forced to consider the cultural, political, and technical systems within which they operated. They were also faced with new forms of beverage containers—in this case, aluminum cans. The troubles that Tomra ran into when simultaneously tackling new markets and new materials indicate the inherent challenges of transferring a set of technologies and organizational solutions between different nations, markets, and cultures.

New Materials, New Affordances

New materials had a significant influence on the beverage container marketplace after World War II. In particular the aluminum can had a strong impact on the production and distribution of beverages on a worldwide basis. Aluminum could not have taken such a central role in modern industry and consumer society without a matching trivialization of the material. "Aluminum was once a precious metal," wrote American Alcoa in a 1969 publication on modern aluminum products.[3] While aluminum is the most common metal on earth, it was not until the mid-nineteenth century that the process of producing it had become sufficiently stable for commercial use. When Col. Thomas Lincoln Casey of the U.S. Army Corps of Engineers installed the one-hundred-ounce (2.8 kilo) aluminum apex on the Washington Monument in 1884, it represented something rare and special.[4] Yet by 1969, aluminum had become one of the most common metals used in industrial production and consumer goods, with almost nine million tons of the material produced worldwide. Aluminum production had doubled in just ten years, enabling widespread commercial use of the metal.[5]

What were the beverage industry's motivations for choosing aluminum cans over reusable bottles? Steel cans for beer had already entered the U.S. market in 1935. Large distributors were particularly keen on switching from glass bottles to steel cans because of cheaper production costs.[6] Aluminum cans were first used commercially in the United States by Coors Brewery, in

1959. Bill Coors, the head of the company, had been disturbed by the sight of steel beer can litter on the roadsides around Colorado. According to Coors biographer Dan Baum, Coors believed that the U.S. government would eventually require a deposit-refund system on steel cans to reduce the littering problem. Coors decided, therefore, that his company should move from steel cans, which were not reusable, to aluminum cans, which were highly recyclable. It took him five years and ten million dollars to develop aluminum can technology appropriate for bottling, but his endeavors quickly proved worthwhile for Coors.[7] In addition to advantages on the littering issue, aluminum cans weighed less and had a better printing surface than steel ones. The U.S. beverage industry quickly followed Coors in moving from steel cans to aluminum ones.

Like the beverage industry, the aluminum industry had an ambiguous relationship with recycling and environmentalism. For instance, large corporations like Alcoa were pro-recycling in the public, yet opposed government legislation requiring it.[8] When aluminum companies began buying back used cans in the late 1960s, they did so because the cans were highly and easily recyclable.[9] Economics dominated their actions. Producing aluminum from bauxite ore has the highest energy use of the processing of any major metal. Recycling aluminum, however, requires 95 percent less energy than producing virgin aluminum.[10] At the same time, recycling aluminum does not degrade the metal, unlike many other materials, making aluminum one of the most recyclable industrial materials. Aluminum producers thus used aluminum buyback programs as a way to decrease the cost of aluminum production. This meant that a recycling program for beverage cans would be extremely attractive for aluminum manufacturers, at least as long as production costs for virgin aluminum remained high.

Despite all these obvious advantages, aluminum cans came to play a highly different role in different markets. As we saw in Chapter 2, the Norwegian government had practically banned disposable containers like aluminum cans through the high taxes initiated in the 1970s. Thus, aluminum cans were virtually nonexistent in the Norwegian market until the late 1990s. In the United States, however, aluminum cans were in widespread use by the late 1970s, and they were also increasingly common in Sweden. Worldwide, RVM producers, and particularly Tomra, developed new types of machines that could handle these containers, but dealing with these new markets entailed more than just changing the technology. In this chapter we will examine the introduction of aluminum can deposit systems in Sweden and New York—two markets in which Tomra was very active.

The Swedish Returpack System

The first RVMs had come to Norway from Swedish manufacturers in the 1950s; however, by the 1970s this relationship had been reversed. The most advanced RVMs in the world originated in Tomra's small factory in Norway. When the Swedish government instituted a forty-öre tax on all beverage containers on March 1, 1973, Sweden became an important market for Tomra, as we saw in the previous chapter.[11] At this time, the metal container producer Plåtmanufaktur AB (PLM) distributed and sold Tomra RVMs in the Swedish market.[12] PLM wanted to follow their packaging through the entire product lifecycle. PLM had introduced beverage cans to Sweden in 1955 and decided to build an aluminum can factory and begin large-scale production of aluminum cans in 1979, in spite of the container tax, through an alliance with Pripps, Sweden's largest brewery.[13] These containers quickly became very popular. By 1980, Swedish consumers had purchased five hundred million cans. PLM's new factory in Malmö could make nine hundred million aluminum cans every year, though not all were intended for the Swedish market. The domestic consumption numbers rose to 600 million in 1983 and 826 million in 1989.[14]

Because there was no return system in place for empty cans in Sweden, littering became a significant problem. A 1982 study found that beverage containers constituted 10 percent of total littering.[15] In an attempt to counter littering accusations, PLM encouraged recycling of its containers through an information campaign initiated in the late 1970s to motivate the public to recycle aluminum cans voluntarily. The pilot study, which piggybacked on the collection system for used paper and textiles, was limited to the small mid-Swedish town of Varberg, leading many to question the validity of the trial results.[16] The trial information campaign was not able to achieve very high recycling rates, the rate being somewhere between 35 and 60 percent, depending on whom you asked. PLM thought the campaign was more successful; the environmental organizations disagreed. Whatever the actual rate, it was still too low to satisfy the Swedish parliament.

PLM's new aluminum can factory had worried Swedish politicians since the news first broke in 1979. They were particularly concerned by the high energy costs of producing new aluminum cans combined with the lack of a mechanism for collecting empty cans for recycling. This problem would only be exacerbated if aluminum cans took an even larger market share away from glass bottles, which were mostly returned and refilled. So in response to what they considered a growing environmental problem, the Swedish parliament, Riksdagen, passed a law on May 18, 1982, that allowed the use of disposable aluminum cans in Sweden only in conjunction with an organization and a system that could handle the return and recycling of these cans.

The new aluminum can law required industry to recycle a minimum of 75 percent of the cans within three years or face a ban on aluminum containers.[17] Riksdagen had established as early as 1975 that producers had a responsibility for handling their products after disposal.[18] As part of the new system, the old fee on disposable containers was removed and a new, differential environmental tax added.[19] The old fee provided the Swedish government with fifty million kroner annually; the new environmental tax would give some of this back. Riksdagen hoped the differential tax would lead to high recycling rates because the brewery industry would gain financially if more containers were recycled. Sweden was the first country in the world to set up such a system, an environmental legislation model in which the industry was free to choose how to achieve this goal, but where the goal was clearly defined—and enforced—by the government.

Being the largest industrial stakeholder, PLM took a lead in developing the new recycling system. To address the requirements of the new law, PLM established a consortium called Returpack to increase recycling of their products. PLM owned 49 percent of this organization, the Swedish Brewery Association owned 49 percent, and the retail industry owned the last 2 percent.[20]

Returpack began planning a deposit-refund system similar to the one used for glass bottles as the most realistic means for reaching the government's recycling goal. Since Tomra made many of the machines that were used in the glass system, PLM invited it to lend its expertise to the task force in 1981. During the next year, Tore Planke regularly participated in meetings at PLM's headquarters in Malmö, where he actively promoted a deposit-refund system centered on RVMs. His most important tool for persuading Returpack to institute a deposit-refund system for cans was a survey of different deposit markets that he had compiled. Based on this data, he created a graph of the relationship between deposit value and return rate that demonstrated that the higher the deposit value was, the higher the return rate would be, but the curve leveled off when the return rate exceeded 80 percent. Based on the curve, Planke argued for a fifty-öre deposit value (approximately fifteen cents), but the industry feared that such a high deposit would make aluminum cans less competitive, compared with refillable glass bottles. A deposit of twenty-five öre was the highest they would agree to. This translated to a return rate of 70 percent at most on Planke's curve.[21]

Although PLM had already learned that its ongoing information campaigns alone would most likely not reach the desired recycling rate of

75 percent, they did not give up on the approach. Keep Sweden Clean (Håll Sverige Rent)—an organization with many similarities to Keep America Beautiful—had been around since 1962, but only as an awareness campaign initiated by the Swedish Society for Nature Conservation. The organizing committee arranged cleanups and events all over the country, but had to close down the official campaign, because of lack of funding, in 1974.[22] The Swedish government had encouraged a more formal arrangement for Keep Sweden Clean since 1969; littering was a problem that should be taken seriously.[23] Therefore PLM and the Swedish Environmental Protection Agency (Naturvårdsverket) decided to create a more stable funding source for Keep Sweden Clean in 1983. A .5-öre contribution from Returpack for every can sold, earmarked for information campaigns, provided Keep Sweden Clean with a sufficient income to became a proper foundation rather than an ad hoc–funded campaign.

THE BATTLE OVER MACHINES

To test the system design in practice, Returpack ran a trial on Gotland, Sweden's largest island, in 1982. Located in the middle of the Baltic Sea, the island offered an opportunity for a more or less controlled experiment, with few beverage containers entering the market outside the distributors' control. All in all, about forty RVMs were installed during the experiment—far fewer than the number of stores on the island. More than one hundred stores did not install RVMs, and these stores received many more complaints about the can deposit trial run than the stores with machines.[24]

Several RVM manufacturers participated in the trial in order to win a share of the new Swedish market.[25] Tomra was the first company to install their machines on Gotland. These were soon joined by machines from six other vendors: five Swedish manufacturers and the American company Environmental Products Corporation (Envipco). Operating out of Virginia, Envipco was the leader in the American RVM market, which had around twenty producers of RVMs at the time. Tomra and Envipco would soon meet again, competing for the new RVM market that opened up after New York implemented a bottle bill in 1984.

While the Gotland trial went reasonably well, nationwide implementation proved problematic. Aluminum cans had been sold in Sweden for years without a deposit system, and consumers had grown accustomed to throwing away their empty cans. Aluminum cans were clearly considered disposable containers. Changing this mindset was hard. Returpack had to teach consumers how to recycle their cans. Newspaper articles repeated Returpack admonitions that consumers had to bring their empty cans—preferably clean and whole—back to the grocery store and leave them at the designated place, depending on whether the store had chosen to use mechanical or manual handling of cans.[26]

Despite Returpack's advertising campaigns encouraging can returns, the twenty-five-öre deposit was not enough to meet the 75 percent recycling rate requirement. From the introduction of the deposit in March 1984 to the end of the year, Returpack collected a total of 283 million cans, but about 100 million cans were still missing. Aluminum can skeptics argued that this confirmed their misgivings with the new deposit system. The recycling rate was particularly low in the countryside. Since RVMs could not accept damaged cans, nobody bothered to pick up such containers.[27]

In an attempt to stave off the criticism and improve return rates, Returpack decided to increase the deposit to fifty öre in 1985. Since Returpack raised the refund to fifty öre after millions of twenty-five-öre cans were already in circulation, customers received a double refund for their previously purchased cans. This cost Returpack millions.[28] But the change also allowed Returpack to reach the recycling targets and avoid a ban on aluminum cans.

Because PLM was interested in selling more aluminum cans, it needed to support the new deposit system to ensure high return rates. The RVM was integrated into the Returpack system because it encouraged those high return rates. The new aluminum can system thus piggybacked on the established solutions in the glass bottle recycling system, where Tomra was a dominant actor. The RVM provided a way to connect these means and goals in practice and thus helped to change the public understanding of aluminum cans from one-way, disposable containers into recyclable containers that consumers collected and returned in RVMs.

THE CAN-CAN RVM

Due to Tomra's involvement in the task force, the company had started developing a new model for the Swedish market in 1981: the Tomra Can-Can. This RVM could recognize cans based on metal type and shape and featured a number of new technological developments, including a metal detector, full-can detector, and can compactor.[29] The Tomra Can-Can was a very compact machine, revealing a kinship with the Tomra Junior. It was designed to fit in small stores, being only 40 centimeters wide and 160 centimeters tall. A small hole on the right accepted empty cans and another hole under this one returned rejected cans. A futuristic-looking LED display showed a welcome message and the total number of cans deposited. All in all, the outside of the machine looked sleek and polished in a way that earlier Tomra machines did not. The inside was similarly advanced and efficient. The machine

Figure 9. Tore Planke and industrial designer Roy Tandberg receiving the Norwegian Design Council Mark for Good Design for the Tomra Can-Can in 1984. Courtesy of Norsk Designråd under Creative Commons Attribution–Share Alike 3.0 Norway license.

compressed the deposited cans and stored them in a container that could take five hundred to seven hundred cans, weighing approximately fifteen kilos.

Tomra's Can-Can machine signaled a shift in the company's attention to the aesthetic qualities of the machine. This was the first time Tomra used industrial designers in the design process. In designing the Can-Can, Tomra cooperated closely through the entire product development process with the design company Nils J. Tvengsberg, in particular the designer Roy Håvard Tandberg.[30] Tandberg had worked with the Ford Motor Company during his education and with Volvo later in his career. By turning to professional industrial designers, Tomra took steps to professionalize its development process.[31]

The design of the Can-Can was very different from the Tomra SP, and miles away from Tomra's first generation RVM. Tomra's machines had traditionally been made of steel or fiber plates. Tomra wanted to give the Can-Can a softer and more contemporary expression.[32] Tomra also wanted to make an ergonomic machine that was easy to use at high speeds. Finally, the machines had to be easy to maintain and to assemble.

The Norwegian Council of Industrial Design's (Rådet for Industridesign) gave Tandberg's Can-Can design its Mark for Good Design award in April 1984 on the basis of "its concept, its functionality and reliability as well as the good overall impression."[33] The council's intention behind this award was to increase the awareness of design as a competitive factor for Norwegian businesses.[34] It wanted to ensure that design became an integral part of product development and functionality, not something that is added as fluff.

The design of the Can-Can also marked a change in Tomra's relationship with technology. No longer just a mundane object in grocery stores, the RVM became an object with consumer appeal. Tomra replaced the metal-plated front of the first generation RVMs with a slick white cover on the SP and a dark plastic cover on the Can-Can. This began an increasing concern with how both grocers and the consumers returning bottles—the technology's actual users—saw the machine.

A fancy design alone could not give Tomra the necessary edge over its competitors, however. During the Gotland trial run, Tomra discovered that PLM had started developing its own RVM as soon as it realized that a deposit seemed to be the only way to reach the necessary recycling rates. To counter this threat, Tomra mobilized its contacts in the grocery industry to make sure that PLM would not corner the market. Since Tomra had sold their RVMs to Swedish grocers since 1973, it had an extensive network of contacts in 1982. They convinced ICA, Sweden's largest supermarket chain, that it was not in the interests of ICA or the rest of the grocery industry to have only one equipment vendor, without competition. ICA and the rest of the grocery industry used their influence to ensure that the new system was open for multiple companies, including Tomra.[35]

But the machine vendors faced additional problems. Returpack originally told equipment vendors that it would buy the RVMs in bulk and lease them to the retail industry; each company then submitted a price offer based on selling its machine directly to Returpack. This requirement was also specified in the original government proposition on aluminum can recycling.[36] In what the Planke brothers later characterized as a "dirty trick," Returpack did not actually buy the machines from the vendors. Instead, it announced that the vendors would sell the machines directly to individual grocers at the same price they had offered Returpack with the quantity discount. This forced the machine vendors, including Tomra, to supply the machines with almost no profit margin and no budget for marketing.

Furthermore, the Swedish government required all the machines sold in Sweden to be manufactured in that country, giving PLM a distinct edge in the market.[37] Tomra responded to this requirement by giving the Can-Can a modular design; it could produce the modules in Norway and assemble the final machine in Sweden. The preassembled components only needed to be set into the cabinet and fastened tight to the frame. This technique enabled fast and inexpensive final assembly and also made it easier for Tomra to use subcontractors in the production process.[38] Tomra thus embedded the politics behind the Swedish requirements in its machine design.

Despite the many challenges along the way, Tomra ended up with the largest piece of the new Swedish market. Although the Tomra machines were the highest-priced machines of all the bids, it received orders for two thousand machines—60 percent of the total number of machines sold in Sweden.[39] Tomra's good reputation among the grocers and the company's market dominance in the glass bottle recycling system very likely contributed to these orders.

Although Tomra had a long time to fine-tune the Can-Can, it faced technical problems in the national start-up. When the majority of these machines were turned on the same day, havoc ensued—Petter Planke called it "1,800 catastrophes," while the 1984 annual report more soberly referred to "certain start-up problems that were due to exceptionally tight time constraints in the product development phase."[40] The consumers were not very happy with the machines, which had been designed with the cheapest solutions possible in order to meet the price range Returpack demanded. By using the modular-design solution, Tomra also lost some control with the final build quality of the machines. Many machines suffered from frequent mechanical breakdowns or had problems recognizing cans; Swedish newspapers were full of articles featuring angry consumers cursing Tomra's "crappy machines."[41] During the next year Tomra had to redesign and replace many machine components, all over Sweden, at a cost of between five million and eight million kroner. In the end Tomra probably lost money on the machines, but it won the market. This would be important for later revenues in maintenance and upgrade contracts, not to mention the brand name awareness it gained.

Tomra succeeded in securing a significant share of the new Swedish aluminum can recycling system. Through a clever technological solution (the modular design) and mobilization of social resources (contacts in the grocer industry), Tomra managed to sidestep Returpack's attempts to give PLM's reverse vending machines a market advantage. The company had already built a solid reputation among Swedish grocers through its their machines for receiving empty glass bottles. Tomra capitalized on this established relationship in the Returpack discussion and thus gained an advantage over PLM, which had few contacts in the grocery industry. This situated Tomra's RVM as the dominant machine in the Swedish Returpack system. What would happen when the company tried the same thing in the United States?

Tomra and the American Market

With the Scandinavian RVM market completely under control by 1984, Tomra's next goal was to conquer the American market. This was, however, not the first time Tomra attempted to sell their RVM in the United States. As early as 1973, Tomra had entered this playing field as a very young and immature company. At that time, only Vermont and Oregon had implemented bottle bills, but many states had some kind of industry-run deposit system in place to handle reusable bottles.

The Planke brothers contacted the Norwegian Export Council in 1973 to help them find international distributors. The Export Council introduced them to Kjell Hvil Pettersen, a Norwegian working on Wall Street. Pettersen arranged a large American publicity stunt for the Planke brothers: they presented the RVM at the Waldorf Astoria in Manhattan. Dr. Frank Fields (NBC's Science Editor) held a two-minute interview with Tore Planke, which was broadcast coast to coast on the six o'clock news. This program presented Tomra's RVM as an economical solution to the resource and littering problem caused by disposable bottles. After the press conference, Pettersen and the Planke brothers toured the East Coast, visiting the headquarters of several supermarket chains to demonstrate the RVM. Pettersen then started his own company, Norwegian Industries, to distribute Tomra machines in the United States.[42]

Tomra met several competitors when they first arrived in the United States. The inventor Samuel J. Gurewitz had installed several hundred bottle return machines on the East Coast. He patented his first bottle return machine in 1956, and appears to have had some success in selling them.[43] He was still active twenty years after filing his first patent, but went out of business when Coca-Cola launched its one-way, disposable glass bottle. This bottle was almost identical to the older, returnable bottle. The mechanical recognition in Gurewitz's machine could not distinguish between bottles with deposits and bottles without. This proved expensive for the grocers who had installed his machine and who had to pay refunds both for deposit and nondeposit bottles, and Gurewitz's business never recovered from this failure. This experience made it clear that U.S. bottle machines needed to be able to discern the nuances of different bottles. The first generation of optical recognition–based RVMs, in particular the Tomra I, was well suited for this, since optical recognition could pick up small differences that mechanical recognition was unable to see.

Despite the obvious market need, Tomra's U.S. installations did not go that well in the beginning. The Planke brothers soon grew dissatisfied with Pettersen, who they found did not have the resources or abilities to sell such an advanced technical product as the RVM. For him, Tomra was a way to get rich quick. The Planke brothers, on the other hand, aimed to make more long-term plans, and began looking for an established distributor with more resources and a larger network.

TEAMING UP WITH MEAD INDUSTRIES

Their chance arrived at the Supermarket Industries Trade Fair in Dallas in May 1973.[44] This was—and still is—one of the world's largest trade fairs for the supermarket and food industries. The Planke brothers wanted to display the Tomra I there, but discovered during their East Coast tour that what American grocers really needed was a RVM that could recognize multipacks of six to twelve bottles (usually six). Most bottles at the time were sold in cardboard multipacks and returned in the same pack. None of the American RVM producers could handle multipacks in a reliable way. After two months of intense work, Tore Planke finished a prototype machine called the Tomra Multimat, which generated much interest at the fair. Modifying the RVM to fit into local conditions served them well, just as it did the following year, when Tomra adapted the Tomra I to work with the Swedish Wine Monopoly conveyor belts.

Mead Industries, the largest paper and packaging corporation in the United States, immediately put in an order for ten machines. Founded in 1846, Mead had pioneered the development of the six-pack cardboard bottle carrier—the multipack. Tore Planke spent two months in Atlanta, where he continued adapting the machine to local conditions and requirements. Mead installed these Multimats in Kroger supermarkets in Atlanta, where Mead was headquartered. Tomra soon entered into a more extensive cooperation with Mead Industries. It thus made sense for Mead to support infrastructure that would encourage the consumption of multipacks. As we remember from Chapter 4, Tomra had already worked with Mead in France, where Mead's local subsidiary distributed Tomra machines.

New orders followed soon after Mead began selling the Multimat in 1973, but around 1977 the Planke brothers discovered that Mead was developing its own RVM in competition with Tomra. Mead claimed that the Tomra machine had too many shortcomings and that they were redesigning it from the ground up, without violating Tomra's patents. However, Mead used the same photocell technology in its machines as the first-generation Tomra machines, a technology that Tore Planke had patented. But to prove that this was the case, Tomra would have to go to court, facing a huge American company that could afford to spend millions of dollars on lawyers. Tomra did not have the time or resources to do this. Neither, the Planke brothers decided, did they need to. As we remember from the preceding chapter, the development of the new Tomra SP was almost complete at this time. Mead based its RVM on old technology that Tomra was about to make obsolete. So instead, the Planke brothers simply wished Mead good luck and canceled all their agreements.[45] Cooperation with the French Mead subsidiary continued, but Tomra pulled out of the American market. As it was, one-way containers were rapidly increasing their U.S. market share in the 1970s. Unless more American states introduced deposit laws, the market for RVMs would decrease. For the time being, the Planke brothers judged it best to wait and see how the American market developed.

Through their first efforts to sell the RVM in the United States in the 1970s, Tomra learned much about the American market and food industry. There was clearly interest in and a market for the machine, but Tomra did not have the organization or network to sell and service significant volumes. Trying to sell their RVM in the United States cost Tomra time, effort, and money, but ended up being mostly an experiment for the company.

REENTERING THE AMERICAN MARKET WITH THE CAN-CAN

Tomra's success with its new high-tech products, the Tomra SP and the Tomra Can-Can, in Europe inspired the Planke brothers to make a new and more ambitious attempt at conquering the American market in the beginning of the 1980s. After Tomra withdrew from the United States in the late 1970s, the American beverage container situation continued to change. Most important, cans began replacing bottles all over the United States. Furthermore, several U.S. states enacted bottle bills requiring the recycling of bottles and cans, opening up new markets for RVMs.[46]

Vermont had passed a bottle bill, which banned nonrefillable bottles but did not require a deposit, as early as 1953. This bill expired after only four years, in large part because of heavy lobbying from the beer industry. The first governmentally mandated deposit beverage system in North America started in British Columbia, Canada, in 1970. In 1971 Oregon implemented a similar system, which outlawed pop-top cans and required a five-cent deposit on most bottles.[47] Vermont followed suit by reintroducing a bottle bill in 1973, this time with a five-cent deposit on most bottles. These early bottle bills were massively contested by the beverage industry. The Oregon bottle bill was one of the most intensely lobbied bills in the history of the Oregon legislature. A 1971 Newsweek issue quoted Oregon attorney general Lee Johnson as saying, "I have never seen as much pressure exerted by so many vested interests against a single bill," but the bill still passed with considerable support.[48] Supposedly, the aggressive and condescending behavior of the beverage industry lobbyists offended many Oregon legislators, who also were influenced by the growing public interest in environmental issues.[49] It is also significant that both Vermont and Oregon were small and had a strong environmental constituency. By the early 1980s, Connecticut, Delaware, Iowa, Massachusetts, Maine, Michigan, and New York had all introduced bottle bills and beverage container deposits. Bottle bill proponents in other states were defeated by increasingly well-funded and well-organized industry opposition, particularly from groups like Keep America Beautiful, which advocated only voluntary recycling programs.[50] In the United States, bottle bills were the result of negotiations between environmental organizations, policy makers, and business interests to create mutually acceptable solid waste disposal policies. Although grassroots environmental organizations were involved, business interests were more important in shaping recycling policies, "although far less socially and politically visible."[51]

The bottle bills were just some of many similar American recycling programs initiated in the 1970s and 1980s. The belief in an imminent landfill crisis was the background for this surge—as illustrated by the near-mythical story of the Mobro waste barge, a small-scale entrepreneurial scheme of transporting garbage that went awry when the barge was not allowed to land in any of the American states where it attempted to bring the waste.[52] People did not want it in their backyard. The introduction of Subtitle D of the Resource Conservation and Recovery Act in 1991 cemented this trend by forcing half of all American landfills to close because of noncompliance with the new regulations.[53] Landfills, just like littering of empty bottles of cans, have high social visibility. Frank Ackerman notes how "recycling was therefore considered critical to avoid the huge expense and environmental burden of additional landfill construction."[54] While the landfill crisis has not been manifested, consumer recycling still remains a powerful tool in modern waste management.

The temptation of the American market proved hard to resist for Tomra. Unlike the situation in the 1970s, when U.S. breweries abandoned their deposit-based return systems in favor of disposable containers, the new bottle bills promised a more stable market for producers of RVMs. Full of confidence from their political and financial success in Scandinavia, Tomra used the Can-Can as the basis for its second attempt to become the leading RVM producer in the American market.

MARKETING THE CAN-CAN

This time Tomra's plans were far more ambitious than in the 1970s. The Planke brothers believed that Tomra now had the business expertise to handle American recycling issues. It was certainly a much more mature company. The Planke brothers had gained valuable experience in Sweden and had built a very solid home base for their company in the European market. Tomra was listed on the Norwegian stock exchange in 1984, and the value of the company was rapidly increasing. The Planke brothers were confident about Tomra's future, and the American market promised to take Tomra to new highs.

The American beverage container recycling systems required Tomra to modify the Can-Can design. In the U.S. deposit states, the individual bottling companies were responsible for collecting the deposits and refunding them to the consumer. The RVM would thus have to be able to sort and register the cans by manufacturer. Tomra modified the Can-Can to fit these requirements by adding a laser scanner to enable barcode recognition. The American version, called the Can-Can Wide, was bigger than the Swedish one. Tomra made this change so that the machines could store more containers and also because American grocery stores generally did not have the space limitations of Scandinavian stores. When Tomra later introduced the bigger Can-Can to Sweden, the Swedes dubbed it the "Tomra Jumbo."

When Tomra demonstrated the new Can-Can at the Dallas trade fair in 1984, the RVM outperformed all the American machines.[55] Whereas the other machines not only required the cans to be clean, but also carefully deposited on a conveyor belt, the Tomra Can-Can could accept, scan, and identify the cans as fast as users could insert them through the opening, no matter how dirty and scratched the cans were.[56] While they had some American competitors, including Envipco, the other RVMs were generally not as technically sophisticated.[57] The Envipco machine—the Can Redeemer—could compete on features, but was bulky and very expensive. The many deposit laws introduced in the late 1970s and early 1980s inspired a good number of companies to seek technological solutions to the container-handling problem; however, few had the know-how to survive in this business. When Tomra made its second attempt to enter the American market, it had twenty-five competitors; by 1986, there were only three. While this certainly had to do with the advanced technology required for reliable container recognition, the Planke brothers learned the hard way that technology was only a small part of the American beverage container problem.

Tomra's efforts to enter the American market signaled a closer attention to marketing than previously. The company's attempts to integrate the Can-Can into the American beverage container recycling system extended beyond technical modifications. In 1983 and 1984, Tomra spent eight million kroner on marketing—an unprecedented amount for the company.[58] As part of the PR work in the American market, Tomra made a marketing video in which Tore Planke demonstrated the features of the Can-Can. He addressed the American audience directly, by highlighting features such as speed, accuracy, flexibility, compact size, safety, reliability, economic control, ease of service, and software control.[59] The video demonstrates that Tomra still thought of the RVM as a high-tech infrastructural machine and that American grocers and supermarkets were its main target group. The company's goal in the advertising campaign was to "move from a position of relative anonymity to gain recognition as the leading supplier of reverse vending equipment."[60]

To fill the anticipated large orders, Tomra invested heavily in its production facilities, opening a new three-thousand-square-meter manufacturing company in Oldenzaal, the Netherlands, in December 1983 and doubled the size of its Asker factory to thirty-four hundred square meters during 1983. Tomra established a subsidiary, Tomra U.S. Inc., in Atlanta in August 1983, as well as a subsidiary in Canada.[61] Through this dramatic expansion, the Planke brothers positioned Tomra to grow rapidly. The company stated that the goal was to grow 35 percent every year and to have a capital ownership of at least 50 percent.[62] Tomra felt secure in this growth strategy. The Tomra SP performed exceedingly well after the company had addressed the initial problems from the Swedish market. Tomra's financial situation seemed equally solid.

The intense marketing effort gave Tomra a contract with Continental Can Company—the world's largest producer of cans at the time—for the sale of two hundred RVMs in 1984. This company wanted to use the machines in a national program based on voluntary return of aluminum cans. The value of scrap aluminum had doubled in the past year, making can recycling in the nondeposit states profitable for Continental Can.[63] The machines were to be tested in the fall of 1984, and if the tests were satisfactory, the agreement stipulated a *minimum* purchase of nine thousand RVMs in the period 1985–1989, at a value of two hundred million kroner.[64] Just like PLM and Mead, Continental Can wanted to follow its cans through the entire product life cycle. This was a major financial windfall for Tomra, as the order represented more than half the total amount of RVMs it had produced so far. Such a large order would dwarf all Tomra's previous successes.

In short, the future looked bright for Tomra. In May 1984, the Norwegian press reported that the company was "carefully optimistic" about the new contract. It planned a growth of 50 to 60 percent for 1985 and 1986. Several newspaper articles called attention to the fact that more and more markets introduced bottle bills, creating a need for the machines Tomra could provide.[65] The potential size of the U.S. market presented an irresistible lure for Tomra, who staked everything on this effort.

THE NEW YORK RETURNABLE CONTAINER LAW OF 1982

Twenty years after the Norwegian comedian Carsten Byhring wondered what to do with his empty nondeposit bottles, Robert S. Winkler—formerly of the deposit state of Connecticut, but now living in Westport, in upstate New York—faced the same moral and logistical quandary. After moving to New York, Winkler's apartment had quickly filled up with empty bottles. Being conditioned to returning empty bottles, he could not bring himself to simply throw them out. "Unreturned bottles and cans had been taking up my valuable counter space for months. They crowded me out whenever I tried to wash dishes, protesting my efforts at cleanliness with annoying clinks and clanks, often falling with a bang on the counter, in which case I would quickly make sure they weren't cracked or broken (don't ask me why). They took up room where pots and pans should have lived undisturbed. They rudely elbowed my toaster oven." One winter night, Winkler counted all his empty bottles and came to the conclusion that he had been "putting up with the trouble, the guilt, the fear of being discovered—all for one dollar and forty-five cents." After this epiphany, he "learned to throw out the empty bottle with little or no remorse." He wondered why his New York friends would want a bottle bill, since it would be "a government intrusion on their lives, a threat to their individualism, but worst of all, a monumental inconvenience."[66]

Winkler shared his concerns in a 1981 letter to *New York Times*, just one of many published in that newspaper that would attack the idea of bottle bills during the 1980s. A majority of the letter writers concluded that disposable containers were preferable to deposit containers for a variety of reasons, among them convenience, environmental friendliness, and economics. It is, however, hard to judge the authenticity of these letters, considering that the beverage industry has often been accused of launching smear campaigns against bottle bills.[67] Some of the letters are certainly extremely compelling and well written. Did they simply represent the opinions of a well-educated and articulate group of New Yorkers, or were they part of a coordinated campaign from bottle bill opponents? We can't say for certain. But it is clear that the bottle bill was controversial.

The New York Senate passed the New York Returnable Container Law in 1982, to take effect on September 12, 1983. As with earlier bottle bill debates in other states, the deposit was fiercely contested in the New York state legislative sessions.[68] New York politicians argued that the bottle bill would reduce litter, save solid waste collection costs, conserve energy and natural resources, and create employment "without placing any substantial burdens on manufacturers, retailers or consumers."[69] The belief in an imminent landfill crisis strongly influenced their opinions. Political support was strong enough to pass the new law in the face of industry opposition. However, the beverage industry managed to limit the bottle bill's efficacy and caused serious problems for an RVM solution to container recycling.

The bottle bill had a direct impact on New York's citizens and grocers. First, beer became more expensive. Distributors raised their prices up to 10 percent because of the extra handling required in collecting bottles and bringing them to recycling centers. Furious retailers accused breweries of using the bottle bill to increase their profits. For instance, both Budweiser and Miller increased the cost of a six-pack by seventy cents after the introduction of the bottle bill.[70]

Second, the bottle bill changed the selection of beverages available to consumers. In particular, New York City grocery stores, which often were small, had to winnow down their selection of beverage brands to make room for empty containers. In most cases, this meant that stores got rid of everything but the best-selling domestic beverages.[71]

Third, the bottle bill pushed breweries to use even more disposable bottles. Because breweries had already removed the infrastructure necessary to use refillable bottles, they faced large expenses in returning to refillable bottles. For instance, Miller Brewing Company—which was the only brewery left in New York State at the time—claimed that bottle washers and other necessary machinery would cost them one hundred million dollars.[72] Because breweries were unwilling to make such investments, the number of disposable containers continued to rise. Grocery stores generally did not mind this development, since cans were easier to stack than bottles.

While numerous New Yorkers opposed the bottle bill, many of the city's less affluent citizens joined environmentalists in embracing it, as empty beverage containers represented a new source of income to them.[73] However, having this demographic group bring empty bottles into stores further antagonized large groups of grocers. Class and race thus became mixed up in the new deposit-refund system, as the people who returned bottles most often were homeless and black. Combined with the inherent messiness of empty bottles, many grocers felt that this did not make for orderly stores. Many stores placed their RVMs outside, but uncleaned bottles could attract roaches and rodents to the store.[74] The manager of the A&P at Fifty-fifth Street and Ninth Avenue said that he was "forced to contend with mountains of dirty, smelly cans that he must store in a basement, angry customers who do not like the disheveled scavengers in the store and outraged neighboring merchants who say the crowds of redeemers outside ruin their business."[75] The grocers thoroughly resented having to receive the empty containers, saying "We're selling goods, not buying trash."[76] The New York State Food Merchants Association argued that the proposed New York bottle bill turned food merchants into "garbage handlers" and shifted the costs of municipal waste management to the private sector.[77]

In short, the small and crowded grocery stores in New York City suddenly needed a way to handle their empty bottles and cans as a result of the bottle bill. To RVM producers like Tomra, the New York situation thus appeared to be quite similar to that of the European markets in which Tomra enjoyed some success.

Indeed, many stores began using RVMs soon after the law went into effect. The *New York Times* published an article on RVMs in 1983, in which it described Envipco's Can Redeemer as a "vending machine in reverse." Envipco's RVMs used laser scanners to read the bar code on the can to determine the brand of the can, what kind of material it was made of, and whether it was empty. And then the machine could crush the can in 2.5 seconds. When the machine had collected eighteen hundred crushed containers, Envipco would return the cans to the distributor, which would pay 5 cents for each can to Envipco. The company thus did not make any money on handling, but instead counted on renting out the machine to the stores for about $145 a month. Envipco planned to ship eight hundred RVMs to New York City in 1983.[78] Five years later, they had about eighteen hundred machines in use in Connecticut and New York, handling about sixty-one million containers annually. This averages to less than one hundred containers per machine per day, which is not all that much, considering that Tomra machines in Norway could handle thousands of containers in a single day.

Envipco was not alone in believing that there was a huge potential market for RVMs in New York. Metropolitan Mining, a company founded in 1983 to recycle aluminum in New York City ordered five hundred Tomra RVMs bound for New York grocery stores.[79] Metropolitan Mining believed that the bottle bill requirements would impel grocers to buy the Can-Can machine. Metropolitan Mining wanted to manage the entire container-handling process, from the grocery store to the recycling plant. However, the newly founded company did not actually have the funds to pay for the machines from Tomra up front—it was hedging its bets on the demand from grocers—and Tomra's future was riding on this bet.

FACING A HOSTILE BUSINESS ENVIRONMENT

Tomra's business concept from the beginning had been to make recycling convenient for consumers, grocers, brewers, and distributors. These groups had a mostly cooperative relationship in Norway, but this was not the case in the United States. While New York bottlers and brewers had not been able to avoid the bottle bill altogether, they did their best to sabotage it. New York consumers certainly had a hard time when trying to recycle their containers. Since the bottlers and brewers had to give a refund for only the cans they had produced, cans had to be sorted not only by material, but also by brand.[80] This required grocery stores to have separate containers for the different brands. Needless to say, this arrangement was very time and space consuming.

Since the New York bottle bill—like those in many other American deposit states—was set up to let bottlers and brewers keep all deposits that were not refunded, the incentive system did not encourage them to work for high recycling rates. In 1985, more than 5.5 billion beverage containers were sold, equaling $275 million in deposits. About $60 million of these went unclaimed, a feature that had been designed into the system. Supposedly, legislators and beverage distributors had reached an informal agreement when the bill was passed that the unclaimed deposits would help cover the costs of setting up the bottle return network. Such a system encouraged bottlers and distributors to make it as difficult as possible to return containers because they would end up making money on those that were not returned. On the other hand, distributors accused the state of not supporting the recycling infrastructure financially, or even giving guidance for how to set up such a recycling system.[81]

At the time Tomra did not realize the political implications of the New York system and attempted to create a technological solution to a logistical problem. This was, after all, where its primary expertise lay. Tore Planke and his engineers had worked hard to modify the Can-Can to the American market.[82] It could read the bar code on cans to identify the brand and producer. The cans were then crushed, compressed, and collected in a single container. A data unit kept track of how many cans were collected from each producer. But as the date to install the units in New York crept closer, Coca-Cola demanded that the bar code readings be proved, arguing that it did not trust the machine.[83] In practice this meant that all cans had to be counted by hand. This deflated grocer interest in purchasing Tomra RVMs, so Metropolitan Mining was able to sell less than a third of the Can-Cans—the rest ended up in storage. Because Metropolitan Mining did not have the financial means to pay for the unsold machines, Tomra faced a huge loss.[84] Other RVM producers of course also struggled with the reluctance of grocers to purchase RVMs, but they were not as hopelessly tangled up in precarious funding schemes.

The problems that RVM producers, and particularly Tomra, faced in New York demonstrated the limits of the RVM as a purely infrastructural machine. Because of the lack of a supporting recycling system, the RVMs did not receive anywhere near the return rates that had been achieved in Scandinavia. Although the Can-Can had been designed for the Swedish political environment, it was not adapted to the U.S. market when Tomra transported it there. In other words, the real needs of the U.S. system were not embedded in the machine. The factors that had made the RVM so successful in Scandinavia were not sufficient for success in the United States. The RVMs made recycling convenient; however, U.S. bottlers wanted recycling to be as inconvenient as possible. The machine functioned in the same way, but operated within a different context. By focusing more on the grocers' logistical aspects of bottle recycling and less on the political, Tomra failed to integrate its product into New York environmental policy. The Can-Can was a puzzle piece that fit perfectly on a different puzzle board.

New York taught Tomra about the importance of cooperating with the brewery industry—and that it could not be taken for granted. New York State legislators intended the bottle bill to remedy environmental problems, particularly littering. However, no efforts were made to ensure that authorities, industry, retailers, and equipment vendors would cooperate. The system was set up so that American bottlers made more money the fewer cans were returned. Yet on the whole, New York's beverage container recycling program was reasonably successful. It achieved a redemption rate of 71.6 percent in its first year of operation and 75 percent over the next few years. However, this return rate was not evenly distributed throughout the state; upstate New York had a return rate of 90 percent compared with only 50 percent in New York City. The five-cent deposit was too small to be an effective incentive for the city's relatively affluent dwellers. In addition, most apartments had storage problems because of their small size. By 1990 littering had been reduced by 72 percent, and the total solid waste stream of New York was reduced by over 5 percent by weight and 8 percent by volume.[85]

In Norway and Sweden, the beverage and packaging industries established independent organizations like Returpack and Resirk (which will be discussed in Chapter 7) to handle the management of deposits and refunds because of a system in which container taxes were lower if recycling rates were higher. Since political authorities did not set clear recycling goals in

New York, breweries spent large sums lobbying against and complaining about bottle bills and deposit acts. Bottle bill proponents accused the breweries of ordering research reports that supported their cause and financing smear campaigns.[86] Building an infrastructure system for recycling technology entailed coordinating and negotiating between these different actors. The RVM no longer merely replaced human labor; it made possible entirely new and systemic approaches to bottle handling. In Sweden, the new recycling system for aluminum cans was designed around the machine, not the other way around. This proved to be more difficult in the United States than it had been with the establishment of Returpack in Sweden just a few years earlier.

FROM HIGH EXPECTATIONS TO COMPLETE DISASTER

The Planke brothers had succeeded in creating high expectations when listing Tomra on the Oslo Stock Exchange in January 1985, at a market value of five hundred million kroner. This made Tomra one of the thirty most valuable companies listed on the exchange.[87] In the mid-1980s, Tomra became one of the most popular of the new high-technology companies in mutual funds.[88] An investor recommendation stated that "the increasing worldwide interest for environmental protection, energy conservation, and resource use makes the company's products increasingly relevant," and valued the existing market potential at no less than eight billion kroner.[89] The promised large contracts with Metropolitan Mining in New York served to increase this belief.

But when Metropolitan Mining could not install the ordered RVMs, Tomra's finances suffered catastrophically. At first, though, the ever-optimistic Planke brothers believed that their problems were just temporary setbacks. When the financial results for the first half of 1985 were not as good as they had promised, Petter Planke explained the poor results as a planned deficit for the first half of the year, while the profits would come toward the end of the year when the large deliveries started. "The growth in 1985 will be the largest in Tomra's history. . . . Beginning this fall, we can reap the benefits of our previous investments," he promised.[90] Petter Planke compared what was happening with what the company previously had gone through when developing the SP.[91] By drawing parallels to previous experiences, the Planke brothers assumed that they would pull through again. However, investors did not share this confidence when the promised profits from the United States did not materialize. As a result, Tomra's share value plummeted at the beginning of 1986.

Tomra's ambition to conquer the American market hit a wall in 1986. The price of scrap aluminum suddenly dropped when the Soviet Union unexpectedly dumped millions of tons of the metal on the worldwide market.

In just four months, the price plummeted by over 60 percent. The interest for can recycling in nondeposit states all but evaporated. Petter Planke asserted that the low aluminum price had little impact on the deposit states, where they would now concentrate their efforts.[92] While this probably was true, Tomra still lost the huge contract that was the basis of their U.S. campaign: a group of private investors bought Continental Can and decided that recycling cans in nondeposit states was no longer in their financial interest. The new owners wanted to focus on their core business—producing cans—and canceled the contract with Tomra. This left Tomra with massive capital investments in new production facilities that would now go unused. The financial result was devastating. Tomra reported a drop in sales from 126 million kroner in 1985 to 77 million in 1986. The end result was a loss of 33 million kroner plus 91 million in extraordinary losses in 1986. Within a few weeks the scope of Tomra's problems became clearer. "We have to admit that we partly underestimated the complexity of Tomra's large-scale internationalization program," stated Petter Planke in February 1986.[93] In fact, the plans for American expansion were a disaster.

The Planke brothers had been blinded by the enormous potential market in the United States and were tempted to take too many large risks. There was an implicit assumption that Tomra could make everything work out as long as the technology was good enough. However, despite the company's extensive technological competence, it could not integrate its machine into the American recycling market. When the U.S. market failed to materialize, the risks that Tomra had taken were exposed. The company simply collapsed and Tomra had to leave the American market.

Integrating the RVM in Aluminum Can Markets

This chapter has examined what happened when Tomra's RVM technology moved into new markets. New material inputs like the French abundance of bottle types had previously led Tomra to design new machines. When PLM introduced aluminum cans to the Swedish market and invited Tomra to help with planning the new recycling system, the new material input required a new machine. In both these cases, Tomra's technologies were successful. The most important reasons for this were that the Tomra RVMs operated within a mostly cooperative business culture and that Swedish and French consumers were accustomed to returning their bottles, so this behavior could with relative ease transfer to cans. Tomra managed to align its technological solutions with dominant business interests. The economical deposit solution carried enough weight to ensure high recycling rates. The Swedish Returpack system thus demonstrated the viability of an industry-organized recycling system centered on the deployment of RVMs to achieve desired return rates. Both the business culture and the consumer recycling culture were highly similar to the Norwegian context in which the RVM was developed. Tomra took this experience to heart, becoming more active in bottle bill politics in the ensuing years. We will return to this theme in Chapter 7.

Tomra's experiences in Sweden and in the United States influenced decisions made in the other market. For instance, Tomra rushed the Can-Can's product development because it needed the order to use as a reference when trying to gain a foothold in the American market.[94] At the same time, Tomra's success with the large Swedish order worked as a mental model for the U.S. efforts. Tomra was sorely challenged by various Swedish business and political interest groups, but managed to overcome these by cleverly implementing the demands of the Swedish market in its technology. The Planke brothers put the entire company's future behind the high-speed product development. However, just being clever was not sufficient when Tomra attempted to repeat the all-out strategy in the United States. The Planke brothers had chosen a high-risk business model for their American subsidiary. If the scrap aluminum price had remained high and if the beverage container recycling systems had been set up to encourage recycling, their plan would probably have succeeded. The market was there and Tomra had a clearly superior technology, but as we have seen, it did not operate in a cooperative business environment in the United States and the Soviet aluminum flood smashed all hopes of entering nondeposit markets.

We have seen how the Tomra machines embodied certain assumptions that in many ways were taken for granted by the Planke brothers because they developed the machine to fit within the Scandinavian system. When the machine moved to the United States, it became clear that these assumptions did not work. Tomra had certain ideas about the users of the machine, but it encountered different cultures, different interest groups, and different power relations in the United States. Their experience demonstrates that while a technology can be easily transferred, moving the cultural context of the machine—in this case, the culture of recycling—is much harder. In the United States, beverage containers, technologies, policy makers, consumers, and the beverage industry formed a sticky web in which it proved hard to integrate the RVM. As a result, the Planke brothers' technological enthusiasm and optimistic growth philosophy led Tomra into serious problems. The American market crushed the company's technology optimism and enthusiasm like an empty aluminum can.

Chapter 6 Greening the RVM

ON THE COVER OF TOMRA'S 1989 annual report, a watercolor of frothing green and blue waves evoked associations with nature and the environment. On the inside pages, more serene watercolors of gardens, flowers, and water provided the background for the text. In the managing director's comments, CEO Svein Jacobsen described Tomra as a "green company" and no longer just a high-tech infrastructure company.[1] According to him, packaging waste was being "drawn into the environmental debate." A four-page section on recycling characterized beverage containers as "an environmental issue."[2] Tomra had clearly taken a new approach to beverage container recycling.

While the RVM today is assumed to have a green connotation because of its role in beverage container recycling, we need to recognize that this was not the case from the beginning. As we have seen in previous chapters, the machine was green only in the sense that it limited littering. The bottle problem had long associations with littering and aesthetic damage to nature. Metal cans and broken glass bottles in the streets and in nature were considered a nuisance. As we have seen in previous chapters, this formed the background for many of the deposit acts that were passed around the world in the 1970s and early 1980s. Yet as we've seen in the story of Tomra's RVM development, the machine was designed as an infrastructural solution for grocers. Although environmental problems had been one of the reasons for Tomra's existence from the beginning, the Planke brothers judged that environmentalism could not be used to market the machine in the early years. This had changed by 1989.

In the late 1980s, both media and environmentalist organizations portrayed green consumers as holding the solution to environmental problems in their shopping bags. A "green wave" of environmental concerns swept over the Western hemisphere, influencing consumers, businesses, and policy makers alike. Tomra wanted to capitalize on this by encouraging these green consumers to fill their shopping bags with empty bottles and cans when going to the grocery store. The Tomra management put considerable effort into establishing a clear connection between Tomra, deposit systems, and environmental protection. Tomra's machines became the method through which environmentally friendly beverage container handling could be achieved.

The new idea of green became an important asset for Tomra, which had always based its products and business strategy on environmental concerns. But earlier concerns were local and centered on littering. The green wave became a new cultural input to the bottle recycling system that enabled RVM producers to take new approaches toward both policy makers and individual consumers. Various forms of business environmentalism, whether called sustainable development, industrial ecology, eco-efficiency, the greening of industry, pragmatic environmentalism, or ecological modernization, became part of Tomra's rhetoric to indicate the "green" disposition of its technologies.[3]

The chapter will analyze the ideological greening of the RVM. Two simultaneous environmental turns took place. First, the RVM was reenvisioned as a technology meeting the needs of environmentally conscious consumers. Second, Tomra as a company changed its profile and production processes in order to develop environmental credibility. As we saw in the preceding chapter, Tomra's expansion into the U.S. market in the mid-1980s turned out to be a disaster for the company. Tomra faced a loss of confidence that had to be countered by organizational and rhetorical changes. Merely producing high-tech machines was not enough for Tomra to remain the world's leading RVM producer. A new, professional management took over the company in 1986, replacing the Planke entrepreneurs. Following the crisis, the new management reshaped Tomra from a high-tech company into a green, environmental business. The chapter will discuss how the company's new management made the environmental aspects of the RVM visible through technology, corporate profile, media, and annual reports. Why was it possible for the company's new management to initiate this greening process in the late 1980s?

The New Management

Until 1986, Tomra was run as a family company by two brothers operating in a classic entrepreneurial mode. Economic historian Joseph Schumpeter described the entrepreneur as a person possessed by the "joy of creating, of getting things done."[4] The Planke brothers fit this characterization well.

They were archetypal entrepreneurs who had lived and breathed Tomra and reverse vending machines for fifteen years. From the early 1970s, they had developed a personal entrepreneurial leadership style that had enabled the company's rapid expansion, but that eventually limited its further growth. Schumpeter recognized that this entrepreneurial phase had to be followed by a new organization if the company was to grow.[5] Genuine entrepreneurs and professional business managers possess very different skill sets and characteristics. The entrepreneur's role is to define and make a product, judge the market for it, and recruit users. Managers, on the other hand, make and maintain systems. Some people are able to do both, as Thomas P. Hughes demonstrated in his studies of system builders.[6] The Planke brothers were not willing—or not able—to make that change in their managerial style.

Every growing organization eventually gets to the point at which the entrepreneur no longer can control or even comprehend all aspects of the company—the so-called entrepreneurial paradox.[7] Tomra came to this point in the mid-1980s. Following the failure in the United States, the company's management was replaced by professional managers with Master of Business Administration degrees. The new management realized that having the "best technology" was not enough, so Tomra began actively influencing the integration of the RVM in society. External events and changes in management style, many relating to the environment and to green consumers, thus transformed the company.

In 1986, analysts in the press trying to make sense of the U.S. disaster concluded that Tomra needed someone who could put on the brakes when the entrepreneurs got carried away; this should have been the task of the executive board, whose members "slept for too long."[8] For this reason, Tomra hired Jan Chr. Opsahl as executive vice president in May 1986. Opsahl had the international and professional background that Tomra needed. He was born in Argentina, studied in English boarding schools, and earned a MBA degree from the University of Strathclyde in Glasgow. Opsahl was hired primarily to lead the negotiations with Tomra's creditors in order to prevent a bankruptcy, but he soon took on a bigger role, replacing Petter Planke as CEO in July 1986.

Opsahl had a completely different leadership style from that of the Planke brothers. The Plankes were playful, open, inclusive, and enthusiastic and tended to believe that all their employees could do a good job if they just found the right place for them in the company. Some thought of them as naive; the brothers themselves thought of their style as being more human.[9] Yet they realized that the company needed a strong leader who could make tough and unpopular decisions. Opsahl fit this bill perfectly. He had a history of turning around struggling companies.[10] A journal even called him "the Rambo of business," probably alluding to his ability to cut away "the dead flesh" of a company in a way that might seem merciless, but that he saw as necessary.[11] The Planke brothers described Opsahl as strong and uncompromising, a contrast to their softer leadership style.[12]

Opsahl quickly took stock of the situation, and he saw that it was bad. The most important cause of Tomra's problems was the company's taking a very high risk in the way it organized the U.S. contracts. In the three large projects that Tomra initialized—constituting 45 percent of the company's total sales—Tomra carried between 50 and 100 percent of the risk.[13] In two of the three projects, Tomra owned all or considerable parts of the companies set up to buy the RVMs. Put simply, the company did not have any real customers in the United States. For example, Tomra signed five-year rental agreements for the 550 RVMs sent to Metropolitan Mining in New York. However, only about 150 of these machines were installed; the rest ended up in storage and Tomra had to pick up the monthly rental cost.[14] Because of these financial guarantees, Tomra faced huge expenses and no U.S. income for years when the market disappeared.

Opsahl wanted Tomra to focus on the core markets of Norway, Sweden, Finland, the Netherlands, France, and Austria, as well increase its presence in Denmark, Germany, and Italy.[15] Tomra knew the market and the culture well in Europe, and the political and business environment was friendlier to beverage container recycling. For the time being, Tomra abandoned all plans of selling RVMs in the United States. Opsahl forbade Tomra employees to even talk about that country. Finally, Opsahl directed the product development group to focus on giving the RVM more advanced functions for users, a new style, and new accessories.

The Green Marketplace

As environmental concerns came to the forefront throughout the world in the 1980s, Tomra wanted to demonstrate a direct connection between the RVM and environmental benefits. In other words, the financial success of Tomra depended on placing the RVM in a global story of environmentalism. How did Tomra create those connections?

Given the increased environmental consciousness that characterized the early 1970s, the Planke brothers did not believe that the world would become a disposable society when they founded Tomra. Even in 1974 they stated this belief: "Worldwide development moves toward more returnable bottles, partly due to new laws, partly through taxes on disposable containers, and partly through regular resource scarcity and environmental concerns."[16] But for this development to continue, they realized that it would have to become so convenient to recycle that there would be no resistance.[17] The Planke brothers believed that technology was the key to creating a society based on the reuse of resources. Yet Petter's experience with the grocery industry had taught him that they could not use an environmental argument to sell a machine to grocers. After all, the Plankes needed to build a successful company that made money. The RVM's job was in the backrooms of grocery stores; thus it had to be given different packaging in order to appeal to their customers.

For this reason, the RVM had a dual identity from the beginning. On the one hand, it was a rational infrastructure machine, receiving and sorting bottles. This was the face that the grocers saw, the machine in the backroom. On the other hand, it was an ideological environmental machine, shaping and reinforcing the values of its users. This was the machine that the Planke brothers hoped to create, a machine that could influence the world in a positive way—and thus expand the market for their technology. In 1992, Tore Planke reiterated this belief, stating that "we have to start with the easy things, the visual pollution. And if we start there, the awareness in the public is getting better with using small efforts. And then, as a second effort, you can go on with bigger things."[18] In this way small, everyday consumer actions such as beverage container recycling became the building blocks of the world of everyday environmentalism.

In the late 1980s, the empty bottles that previously had been a local littering problem became a global environmental problem connected to resource use through larger social discourses. According to the newspapers, new deposit acts were expected around the world, creating a need for Tomra's products—and thus Tomra was riding the environmental wave.[19] From 1987 on, profits were up every year. In 1989 Tomra's CEO Svein Jacobsen attributed this to the growing environmental consciousness in society and the marketplace. Jacobsen later stated that "the green [of the company] was not planned. It was something that came with the green wave."[20]

What did this new notion of environmentalism entail? As we have discussed in previous chapters, consumers, activists, and policy makers all turned their attention toward environmental issues starting in the early 1970s. Two signal events were Earth Day and the European Year of the Environment in 1972. However, it was not until the late 1980s that the green wave finally crested.[21] In just a few years, attention to environmentally friendly consumer behavior dramatically increased. While this was a global development (at least for the affluent parts of the world), Norway was at the forefront.

Lester Brown, director of the World Watch Institute in Washington, DC, declared 1988 as the year when environmental issues became front-page material in newspapers all over the world.[22] This was especially true of Norway, which Brown called an "environmental superpower."[23] What made Brown characterize Norway in this manner?

NORWEGIAN ENVIRONMENTALISM

Nature conservation has a long tradition in Norway as an original environmental concern. Bredo Berntsen's history of Norwegian nature conservation and environmental protection traces the relationship between humans and the Norwegian landscape. He argues that people gradually recognized their responsibility for acting in ways that use nature without consuming it, so that it is preserved for posterity.[24] The Norwegian Society for the Conservation of Nature was founded in 1914 to "raise and maintain the sense and interest in our people for protecting the nation's nature."[25] This early tradition was primarily concerned with creating national parks, restoring "natural" conditions, and preserving wilderness and wild animals, just as it was in the United States and in the rest of Europe.

The new environmental concerns of the late 1980s had a different foundation, being centered on consumer actions, industrial pollution, and a global environment—with consumer actions leading the way. Consumption researchers stated that "nothing has changed buying habits more than the environment."[26] In a 1990 report by Norsk Gallup Institutt A/S, the green consumer was defined as "someone who is very concerned with preserving nature, working against pollution, and increasing environmental protection. At the same time, this is a consumer who buys phosphate-free detergents and recycled paper, and who tries to avoid artificial additives."[27] Recycling, buying green products, and advocating producer responsibility were all seen as ways to help the environment. Small, everyday consumer choices now mattered. However, these choices had to be convenient. "Only enthusiasts travel far to get rid of glass. It must become more convenient to get rid of waste in an environmentally friendly manner," asserted a group of housewives, environmentalists, and sanitation workers who published a report on Oslo's waste in 1989.[28] They concluded that recycling had to start at home, with source separation of household waste. By doing so, they highlighted the central criteria for everyday environmentalism: convenience.

The everyday environmentalists of the late 1980s and early 1990s put pressure on businesses to make environmentally friendly products. Tor Killingland, an officer of the Norwegian Society for the Conservation of Nature, said that "consumers are making more severe demands than earlier.

Now they don't only ask how the product functions. It is no longer only the color and shape that matters. The customers also want to know how the product will function as waste, after its lifetime."[29] Grocers accommodated green consumers by stocking green products and reducing unnecessary packaging.[30] Being green became trendy for young people, who "have become far more environmentally conscious lately, and are increasingly asking how they can behave in ways that spare the environment," said Marit Nyborg, the leader of Nature and Youth (Natur og Ungdom), an environmental youth organization whose membership had doubled, reaching six thousand to seven thousand people, in just a few years.[31] Belief in the green consumer was strong. Dag Hareide, president of the Norwegian Society for the Conservation of Nature, stated that "it is primarily at the supermarket and through waste separation that the average consumer can have a positive influence on the environment."[32] The idea of the green consumer helped marketing agencies and environmentalist organizations find a common cause. "A consumer who buys environmentally labeled goods will feel that he is environmentally friendly, and he wants to know that other people see that he is," according to the advertising agency Olgilvy and Mather Norway.[33]

This changing consumer behavior was part of a larger global trend. The environmental wave that was sweeping Europe and the United States changed the way consumers, businesspeople, and policy makers thought about the environment. In particular, the vague and flexible concept of *sustainable development* became popular and influential. Highlighted in *Our Common Future* (also known as the Brundtland report after the Norwegian, Gro Harlem Brundtland, who led the commission), sustainable development quickly turned into a new paradigm for understanding, or maybe even resolving, the relationship between economic interests and environmental protection.[34] "Economy and ecology have been entered in separate account books. If we are to take the Brundtland commission seriously, these accounts need to come together in a whole," the writer Sidsel Mørck stated in an editorial.[35] However, some Norwegian politicians criticized the report for putting too little emphasis on concrete recycling technologies.[36] The green wave of the 1980s got its driving force from the interplay between green consumers and green business. Everyday environmentalism as a system was dependent of consumer choices—and the technologies that supported these choices.

THE RISE OF GREEN BUSINESS

In the late 1980s and early 1990s, politicians, investors, environmentalists, and consumers actively explored how Norwegian companies could contribute to the environmental trend. At a 1990 conference on environmental technologies arranged by the Royal Norwegian Council for Scientific and Industrial Research (Norges Teknisk-Naturvitenskapelige Forskningsråd [NTNF]), deputy secretary Karin Refsnes of the Ministry of the Environment said that "the market, that is, the purchasers of the businesses' products, is ultimately perhaps the most important driving force in convincing business to switch over to environmentally friendly production."[37] Her argument was supported by Helge Fredriksen, executive director of the Environment and Resource Section in the Confederation of Norwegian Enterprise (Næringslivets Hovedorganisasjon [NHO]), who said that consumers and environmentalists were "deliberately using their power to force industry to take nature and the environment into account."[38] Statements like these reveal how both policy makers and business organizations had begun to acknowledge the influence of consumer environmentalism on business.

Business responded to the rise of the green consumer by commodifying environmentalism. Harald Rensvik, who was director of The Norwegian Pollution Control Authority (Statens Forurensingstilsyn [SFT]), asked for more business initiatives to "sell environmental protection."[39] Many companies followed exactly this strategy by actively portraying themselves in a green light in order to reach consumers who chose to buy products based on their environmental advertisements, packaging, or profiles. The most obvious example is recycled paper, which consumers eagerly requested in the late 1980s.[40] The "eco-labeling" of products is another good example. The Nordic Council of Ministers established the Nordic Swan Label in 1989 as a certification of products' environmental friendliness.[41] In 1990, Germany passed an act that extended producer responsibility for packaging all the way through the product's life cycle. The consortium Duales System Deutschland AG (DSD) instituted a "Green Dot" trademark to certify that a product was part of their recycling system.[42] The European Union established their Flower certification in 1992, extending environmental certification to all the EU.[43]

However, many debated whether green business could really help solve environmental problems. Jan Omdahl, editor of *Bellona Magasin*, asked, "Can green capitalists and green consumers save us from environmental disaster, or do we need a radical change in the world economy?"[44] Bellona was a new and highly visible environmental organization, established in 1986 to work with business instead of having an oppositional attitude.[45] Omdahl argued that a green economy had not found acceptance in Norway. This did not stop several businesses from trying. Knut Utstein Kloster Sr., for example, founded the company Green Business A/S to invest in green technologies and green services. Chairperson Jørgen Randers stated that "we want to show the world that it is possible to make money by engaging in environmentally friendly production." The company wanted to do this by investing in green companies. He predicted that public and consumer "demand for environmentally friendly alternatives to today's goods and services will grow faster than other demand in the 1990s."[46] Kloster also founded the environmental consulting company Miljøkompetanse A/S, again with Randers as chair.[47]

The green wave rescued Tomra after its failed attempt to win the American market. Investors, politicians, and consumers embraced Tomra as an environmentally friendly company. In just two years, it managed to regain its momentum after the 1986 crisis. "The reverse vending machines are a sought-after product all over the world," stated Erik Thorsen, then the marketing director of Tomra.[48] "The belief in deposit systems grows larger and larger both for reusable and disposable containers," Petter Planke followed up in the same interview. Business analysts were confident that the rising environmental consciousness would lead to more deposit laws and thus greater demand for Tomra solutions. "The big bottleneck for Tomra is how soon new deposit systems will be introduced, because with today's environmental focus, few doubt that they will force their way."[49] CEO Jacobsen confirmed that Tomra agreed with this analysis, stating that "every development concerning beverage container deposits in Europe point in Tomra's direction."[50] In Tomra's opinion, the green wave thus would translate directly into increased sales for the company.

Tomra's stock started to be included in green portfolios, among them the British Merlin Ecology Fund and the Norwegian MiljøInvest.[51] These funds selected companies with a clear environmental profile based on the belief that they would do better in the future than would "regular" companies. The demand for green shares was high. Tomra's position as "one of the few green shares available in Europe today" contributed to driving up Tomra's share value.[52] Carlos Joly, the CEO of Skandia, a bank that owned two green portfolios, said that green shares were "particularly attractive. More and more investors want a part of green companies, both to help with environmental protection and because there is reason to believe that the environmental industry will grow especially fast in the next ten years."[53]

Tomra recruited Jørgen Randers to its board of directors in 1991. Randers had earned a PhD in economy from MIT in the early 1970s and was one of the coauthors of the influential book *Limits to Growth*, which modeled future environmental scenarios based on population growth and limited resourses.[54] He had served as president of and professor of policy analysis at the Norwegian School of Management (Handelshøyskolen BI) in Oslo from 1981 to 1989. He then retired as president and took a two-year sabbatical from his faculty position because he "wanted to create something . . . to take something small and make it big."[55] Randers's background afforded him a unique position from which to interact with the greening of business. His work with *Limits to Growth* gave him environmental credibility, while at the same time he had a solid business background. In short, he was perfect for making Tomra's RVM a green machine.

The emergence of a green business discourse in Norway set the stage for a major reenvisioning of the RVM. Since consumers and businesses alike looked for ways to become greener, the bottle recycling that the RVM enabled increasingly took on a green hue. Tomra had long marketed itself as a company that made a hole in the wall for grocers, as a company that solved infrastructure problems. In the second half of the 1980s, investors started to see Tomra as a deliverer of environmental technologies. This development was not just an external influence on the company; Tomra also actively participated in creating an image of its products as green technologies.

Creating the Green RVM

Tomra attempted to find a way to latch on to the green wave of everyday environmentalism—to position the RVM as the technical assistance that environmentally conscious consumers needed for convenience. While most green consumers already recycled their empty bottles, Tomra wanted to reinforce the impression of the RVM as an obligatory point of passage for everyday environmentalists. One way that businesses are able to shape their public image is through their annual reports. At the very minimum, an annual report summarizes a company's financial performance for the previous year. Many annual reports are more thorough documents with a significant narrative structure, interpreting and explaining the year's events and performance as well as informing readers and investors about future plans. They are sophisticated products of a corporate design environment, as part of a company's branding and image-building efforts. Annual reports "communicate implicit beliefs about the organization and its relationship with the surrounding world."[56] By studying Tomra's annual reports, as both textual and material artifacts, we can see how conscious the company was of its position, and how it actively started the greening of the company as part of the rebuilding process after 1986.[57]

Tomra's annual reports were extensive documents highlighting the high-tech nature of its products and how it could solve the infrastructural problems of bottle handling. Until 1982, its annual reports were typewritten documents, but their visual presentation improved gradually as Tomra began using computer tools.[58] The 1986 annual report emphasized the management changes taking place in the company and the strategies the company would follow to rebuild its business. The word *environment* is not used at all in this report, and the term *reuse and recycling* only once.

The 1987 annual report signals a more professional and visually attractive presentation of the company. The report is a glossy twenty-four-page publication, featuring photos of a glass bottle filled with water and covered in condensation (Figure 10). A brushed aluminum look is used on some pages. This use of images creates associations with the beverage containers that passed through the RVMs. Yet it is a clean and modern look, far from giving connotations of waste, trash, and recycling. Two 2-page spreads show landscapes overlaid with a Tron-like computer-generated surface disappearing into the horizon. The text reads, "The future is with Tomra"; "in a world with a growing need for our products, the most important challenge in the year to come will be to maintain our position as the world's leading supplier of used beverage container automation systems."[59]

We see here how Tomra assumed that increased beverage consumption would automatically translate into increased sales of their RVMs. As it did the year before, Tomra strongly emphasized that the 1986 crisis was over: "In 1987 the Tomra Group has once again proved itself to be a viable operation."[60] The environment is hardly mentioned at all, except in the statement that the company was "well prepared to meet environmental challenges with proven products for the return of all types of used beverage containers."[61] Despite the updated look of Tomra's annual reports, it is clear that the corporate profile was still in line with the high-tech image that Tomra had cultivated since the company's inception.

The company's 1988 annual report displays the beginning of a turn toward environmental concerns. This report had some things in common with the previous year's report: it was a glossy twenty-four-page publication and photos of the Tomra Can-Can and the Tomra 300 were used several times. But the document reveals a softer side of Tomra. Gradients in pastel colors formed the background of every page. The last two pages were dedicated to the environment, showing a large photograph of a pristine winter landscape, and presenting a short text, with the heading "A Clean Environment—Our Common Responsibility." The text itself focused on the practical aspects of deposit systems for beverage container recycling and how it was possible "to combine the advantages of new types of packaging with tomorrow's tighter environmental requirements" rather than a broad, moral narrative about environmentalism.[62] Although environmental concerns appear only on these last two pages, we see the beginning of the company's new rhetoric on beverage container legislation.

Figure 10. Cover of the Tomra annual report from 1987 showing a sparkling glass bottle covered in water droplets. The focus on the bottle emphasizes the high-tech nature of Tomra's business and its focus on grocers as the customers. Courtesy of Tomra Systems ASA.

The 1989 annual report showcased the company's "position as one of Europe's 'green' companies."[63] Instead of just hinting at environmental issues, this year's publication integrated environmental concerns as the primary motivation behind Tomra's products. This was the company's first report to be published on recycled paper and featured reproduced watercolors by the artist Egil Torin Næsheim, showing nature, flowers, and the sea (see Figure 11). The various undulating landscapes and seascapes on the paper serve as the background image for every single page in the report, creating an impression of organic environmentalism far beyond the clean and polished technological imagery of the 1987 report.

Figure 11. Cover of the Tomra annual report from 1989 with a partial image of Egil Torin Næsheim's pastel watercolor *Aquarelles*. Tomra's choice of image illustrates the company's green turn at the end of the 1980s. Tomra rode the green wave both in art and business. Courtesy of Tomra Systems ASA.

In the text, Tomra put considerable effort into establishing a clear connection between RVMs, deposit systems, and environmental protection. Svein Jacobsen's "Managing Director's Comments" stated that "problems related to the proper disposal of packaging waste have become the focus of greater attention and to a steadily increasing degree are being drawn into the environmental debate. The dynamic on the packaging side will force the active participation of the authorities and the environmentally conscious. The 1990s will be characterized by a political debate on the recycling of all packaging."[64] Tomra found that as a result of its international market position, it was increasingly "being asked to provide advice when national deposit and recycling schemes are considered. This does not mean that we inflexibly defend deposit schemes as the only environmentally acceptable arrangement."[65] Yet the conclusions from their four-page discussion of recycling options for beverage containers were clear: high deposit charges and convenient delivery arrangements "guarantee a high return percentage for used beverage containers."[66] Tomra created a rhetorical construction of the future in which RVMs were central. As we will see in Chapter 7, the company continued to actively promote such deposit systems.

In 1990 Tomra's annual report showcased the company's association with the "environment." The public had recognized Tomra as a green company, and the company capitalized on this, asserting, "Tomra wants to be perceived as an environmentally friendly company."[67] The report highlighted Tomra's winning the Environmental Technology Award at the 1990 World Action Conference in Bergen for the company's "significant contribution to cost-effective and environmentally friendly deposit-refund systems in Norway and abroad."[68] The last sentence in the report is a quote from *The Economist*: "The environment may turn out to be the biggest opportunity for enterprise and invention the industrial world has seen."[69] Clearly, Tomra now fully embraced a vision of its future as intertwined with environmentalism. Tomra acknowledged that its future growth depended on how the supermarket industry related to global environmental trends.[70]

The publication of the 1991 annual report coincided with the company's twentieth anniversary. The report was bound in raw-looking recycled paper and used the slogan "20 Years for a Better Environment." However, the company devoted more time, space, and attention on deposit systems than on the environment in general this year. In the same way that green consumers gradually readjusted their high environmental ideals in the early 1990s, Tomra abandoned the "soft" focus on environmental consciousness, and it now put more effort into advocating deposit systems. The report featured a four-page country-by-country breakdown of beverage container recycling legislation in Europe and the United States. In so doing, Tomra acknowledged that even if it was providing an attractive environmental product in the midst of the green wave, its business model still depended on mandatory deposit systems to provide the incentive for consumers to recycle.[71] At the same time, Tomra clearly recognized the power of the green consumer. The report concluded with the claim that "the fight for a better environment is steadily becoming a more important part of everyday life."[72]

Because this was Tomra's twentieth anniversary, they also published a small brochure highlighting the company's history and future. This brochure mostly focused on high technology, but ended with a paragraph on the environment: "Today's consumers are becoming increasingly environmentally conscious and prefer supermarkets with an active environmental commitment through products and return systems. Using Tomra's systems, a clean and attractive profile is obtained."[73] We see how Tomra recognized the growing environmental consumer concern and attempted to position the RVM as a way for grocers to reach these consumers.

THE ANNUAL REPORTS REVEAL that after the MBAs took over the leadership of the company in 1986, they put considerable effort into reshaping the corporate image. However, in the years after its twentieth anniversary, Tomra realized that it was not enough to provide an environmentally friendly product; it also had to implement an environmentally friendly production process in order to be a legitimately green company.

Starting in 1991, Tomra began greening its corporate profile by adding a statement on the company's environmental status under the general "organization" heading, explaining, "Tomra wishes to be regarded as an environmentally friendly company. For this reason the Tomra Group's products and manufacturing processes have been based on an environmentally acceptable profile. The company does not pollute the external environment in its activities."[74] In 1994, Tomra completed its ISO 9001 certification. In the annual report, this was tied directly to its environmentally friendly profile.[75] The next year, Tomra incorporated the International Chamber of Commerce's Principles for Environmental Management because "as an environmental company Tomra feels a special responsibility toward the environment, and places considerable emphasis on environmentally acceptable solutions both in manufacturing and on the office side."[76] Tomra carried out internal ecological audits of all the group's activities—this included everything from sorting at source, preferred use of recycled paper, and increased use of e-mail. Tomra stressed that the RVMs were also becoming more environmentally friendly through recyclability marking on all plastic parts, limited energy consumption, and improvements in cleaning kits and maintenance instructions in order to prolong the lifetime of machines. The report specifies that Tomra's activities did not cause pollution, whether of the air or water or through noise or otherwise and that the energy used came from hydroelectric power.[77]

Tomra became more quantitative in its analysis of corporate environmental impacts as the years went on. Its 1995 report outlined efforts to reduce the consumption of nonrenewable resources and the development of recyclable RVM components and quantified the amount of emissions and energy consumption attributable to the manufacturing process.[78] These changes in corporate profile were in line with general financial trends. For example, the financial newspaper *Finansavisen* concluded that environmental concerns were no longer a "soft" concern without interest for profit-maximizing capitalists, and that there was a clear correlation between a firm's financial and environmental performance.[79] The environmental report of 1997 introduced the life cycle analysis concept as "a tool for manufacturing environmentally responsible goods using the best available technology within economically and ecologically defensible limits." Tomra advocated looking at the entire RVM life cycle, from research and development to the recycling of used RVM components.[80] By 1999 the life cycle analysis included measurements of the "global warming potential" of the RVMs as well as energy consumption by specific type of machine. The accounting of emission is broken down by operational unit, including R&D, production and logistics, and installation and service. Through this strategy, Tomra wanted to measure and improve its "eco-efficiency."[81]

Tomra took its environmental accounting one step further in 2000, when it introduced both the direct and indirect environmental impacts and benefits of the RVM: "The environmental savings generated by recycling billions of used beverage containers include savings of energy, reduced raw materials extraction, reduction of litter, and reduction of landfill space. Tomra's indirect impacts are positive to society and the environment and should be increased."[82] Tomra now implied that this recycling could not happen without the hole in the wall. In its view, beverage container recycling systems had become totally dependent on its RVM technologies. Because of Tomra's efforts to green the company, it achieved a top "sustainability score" within the pollution control and waste management group in the Dow Jones Sustainability Group Index, solidifying its position in green share funds.[83]

Throughout the 1990s Tomra created a form of business environmentalism by reconceptualizing its idea of the machine and the latter's place in society. By using tools like life cycle analyses, Tomra extended the reach of the machine throughout society, taking credit for the environmental benefits of all the containers passing through it. Consumers' everyday environmentalism thus not only helped provide empty beverage containers for the machine; it also allowed the machine to take on an environmental identity. The green of consumer actions rubbed off on the machine.

Targeting the Everyday Environmentalist Consumer

When Tomra began greening its corporate profile in the mid-1980s, it was not clear to the company whether its form of business environmentalism would be profitable. As Jacobsen stated, they did not plan the green wave. Instead, consumers changed their idea of recycling, and Tomra followed suit. This is an example of user pressure changing a company. When users moved in a new direction, Tomra followed. At the same time, Tomra had learned from its New York failure that it had to teach consumers to recycle. Tomra's sales director, Svein Dagestad, reasoned in 1986, "The 'use and throw away' mentality is widespread in the U.S., but it is not unchangeable. If we can make the Americans realize the advantages of reuse and recycling, much will be gained."[84] Tomra wanted to build on this knowledge. From the 1990s Tomra wanted to be perceived as a company that helped consumers feel they were doing something for the environment. You can drink your beer and your soda and then you just take your empty containers back to the store and save the environment. Encouraging this behavior was vital for Tomra's business strategy.

In the annual report for 1990, Tomra had encouraged grocers to recognize the strong international focus on the environment, pollution, and resource use—in short, the green wave. "Forward-looking supermarkets," Tomra claimed, "meet this challenge by selling environmentally friendly products in environmentally friendly packaging and profiling their stores with customer-friendly Tomra machines for convenient handling of returns. Far from all supermarkets have cashed in on the opportunity that an environmental focus can be as a competitive factor to draw more customers."[85] Tore Planke reiterated this argument later: "More and more supermarkets in the U.S. draw customers by cultivating a profile centered on environmental protection, health food, and other green trends. Returning empty cans in a can machine makes Americans feel that they are doing something to save the environment, and Americans like that."[86] By equating Tomra machines, environmental protection, happy customers, and increased profits, Tomra encouraged grocers to install the company's RVMs.

At its twentieth anniversary, in 1997, Tomra changed its slogan to "Helping the World Recycle," to reach consumers. As the annual report noted, "The fact that our activities are profitable for our customers as well as Tomra, while helping people take better care of their environment, only make them even more exciting and satisfying."[87] Tomra equated the environmentally friendly qualities of its products with its profitability. Environmental issues became an integral part of the company's products, identity, and goal setting. Yet the 1997 annual report clearly presented Tomra's belief that "those who believe that manufacturers of reverse vending machines can ride an ever-lasting wave of 'green' enthusiasm should rethink his or her [*sic*] position."[88] The company's challenge was that consumers are multifaceted and environmentalist sympathies alone might not be sufficient motivators for increased recycling. Tomra aimed "to make recycling an attractive alternative for everyone. This requires convenience and efficiency in handling returns as well as direct appeals to consumer motivation."[89] To make using its machines more appealing to consumers, Tomra turned to four approaches: improvement of storefronts, coupons, donations of deposit refunds, and holistic recycling systems.

IMPROVING STOREFRONT ENVIRONMENTS

Tomra attempted to encourage people to use the RVM by creating new, enticing, and comprehensive storefront environments. The company commissioned Andersen Consulting to propose a vision of how the store of the future might look: "A Tomra-equipped area figures prominently in this vision because it reflects the need of consumers who increasingly think and act in an environmentally responsible manner."[90] Store recycling environments became increasingly more colorful in the early 1990s, especially in recycling centers specifically targeted at children, featuring recycling characters like dragons, dolphins, kangaroos, and frogs. Tomra recognized that kids were one of its most important—and most challenging—groups involved in its products. Children often collect bottles for a little bit of extra pocket money. They also can be very wily. In the 1990s, Tomra decided to capitalize on this by inviting a class of fifth graders to test their new models. The students had access to all kinds of low-tech equipment and were promised a crate of soda if they found a way to trick the machine—which they of course did. It often took several trials like this before Tomra's engineers managed to make the new model sufficiently tamper proof.[91] The focus on visually attractive storefronts to pique the interest of even the smallest consumers demonstrates Tomra's efforts to create a culturally pervasive recycling system with an RVM in the center.

Tomra became increasingly concerned with the presentation of the RVM in the store environment. One main goal was to demonstrate that the machine was not simply a waste container, but something clean and hygienic. As Chapter 3 showed, many grocers used the hygiene argument against the return of bottles in their stores. Even though the RVM was designed to avoid this problem, a lack of cleaning still caused problems with odor, something that American grocers, in particular, claimed when protesting the introduction of bottle bills.[92] Internal newsletters show that Tomra encouraged grocers to clean and maintain their RVMs more often, as a way of mitigating this problem. Tomra also launched cleaning and maintenance subscription services to make this work more convenient for grocers.[93]

Tomra also reshaped the machine itself to augment the recycling experience for consumers. Because so much of the actual work of the RVM takes place behind the wall, most consumers never see more than the hole in the wall and the plastic cover of the machine in the store. In an attempt to strengthen their appeal to consumers, Tomra had begun improving the visual design of their RVMs, starting with the Can-Can. By including professional designers in the process, Tomra attempted to infuse the meeting of the consumer and the machine with social values. In 1997, environmentally friendly consumers could return their empty bottles and cans—whether they were made of glass, plastic, aluminum, or steel—in the brand-new Tomra 600 RVM. This was the most technically advanced machine Tomra had ever made. All beverage containers could be inserted horizontally through a round hole in the wall, solving the stability problems that arose in the older machines, in which bottles had to be vertical when inserted. Online functionality connected the T-600 to a central computer, enabling monitoring, software upgrades, and data collection over a modem connection. A high-resolution video camera with an electronic light sensor paired with powerful computers scanned the inserted beverage containers at high speed and could identify all sorts of containers accurately. The T-600 RVM was not just a recycling machine; Tomra now explicitly used it as a communication tool targeted at everyday environmentalists.[94]

SHOPPING FOR THE ENVIRONMENT

As part of the new orientation toward consumer-users, Tomra began looking for new incentive structures for consumers in both deposit and nondeposit markets. For instance, in 1996 Tomra experimented with "bonus points" in England for those stores with customer loyalty programs.[95] Tomra also attempted a lottery version of the RVM in Greece and Argentina in 1997—the RVM gave the consumer a scratch-and-win lottery ticket instead of the regular deposit slip.[96] Another, more ambitious experiment was the use of couponing. In the American market, consumers had grown accustomed to using discount coupons when shopping. These coupons give a small discount on specific products or brands. On August 1, 1997, Tomra North America initiated a coupon test in the Michigan market to capitalize on this consumer habit. The RVMs that participated in the experiment had to be upgraded with better printers, more memory, and online capabilities. After customers returned their bottles, a coupon for Coca-Cola, Pepsi, Hershey's, Kraft, or Procter and Gamble was printed on the receipt. Because RVMs are placed at the front of stores in the United States, the customer would receive the incentive before shopping, which possibly influenced purchasing choices.[97]

Tomra regarded coupons and other reward-based systems as "an attractive way to promote environmentally responsive handling of empty beverage containers" in nondeposit markets.[98] By targeting consumers' wallets through financial incentives like deposits, coupons, and lottery tickets, Tomra attempted to influence their behaviors and in some cases, even their values. After all, "the time the consumer spends in front of the reverse vending machines is a golden opportunity to catch his or her attention," as Tomra argued in 1996.[99]

"PRESS YELLOW FOR HOPE"—MARKETING GOODWILL

In the 1990s, some charities hoped to piggyback on the goodwill of green consumers. Tomra designed a machine option in which consumers could press a yellow button when returning containers to donate their deposit to the Salvation Army, Norwegian Refugee Council (Norges Flyktningeråd), and Norwegian People's Aid (Norsk Folkehjelp).[100] Tomra upgraded several hundred RVMs with the yellow donation button in Norway (Figure 12).[101] As the 1996 annual report described it, "You enter the store carrying bottles and cans. Your children make a beeline to the environmental recycling center where the reverse vending machines are installed. In a generous mood after feeding all your used beverage containers into one of Tomra's machines, you donate your refund to a charity simply by pressing the yellow button instead of the green one."[102] The pilot project in 1996 had been successful.

Approximately 5 percent of customers donated their deposit, meaning that each Tomra machine could earn sixty thousand to seventy thousand kroner per year for the charities. "The high marketing value in increased customer loyalty and goodwill for the store will also be a positive outcome for the project."[103] In order to participate, the stores had to upgrade the RVMs to include online functionality.

In spite of high hopes and the successful pilot study, the yellow-button experiment was not a big success in practice, because although most consumers thought it was a good idea, they still put the refund in their own

Figure 12. "Press yellow for hope." Tomra advertisement encouraging consumers to donate their refund to charities, 1996. Courtesy of Tomra Systems ASA.

pockets. In the end, only 3 percent donated the deposit and the RVM never became the money-making machine that the charities were hoping for. One store even had to block the yellow button because many customers pressed the wrong button and demanded their money back when they discovered that they did not get their refund.[104] One customer commented, "The machine itself is invaluable, but I'm not too sure about this donation thing."[105] This is an important clue to what actually matters to consumers when they recycle.

Everyday environmentalism works best when it's not too expensive or inconvenient for the consumer. Furthermore, an important element of charity donations is the acknowledgment and recognition of doing something good. The anonymity of using the RVM, which can be an advantage when returning all the beer bottles after last weekend's party, does not allow the everyday environmentalist to be seen as doing good. This might be why the Refugee Council originally estimated that RVM donations could contribute up to 140 million kroner annually, but at the beginning of 1999, they had only collected 1.5 million kroner.[106] The project was canceled in April 1999, after serious internal conflicts in the Refugee Council.[107] Despite the failure, it shows that Tomra experimented with new ways of presenting the RVM to its users.

DEVELOPING HOLISTIC RECYCLING SYSTEMS

Tomra also attempted to reach users through the development of holistic recycling systems. The U.S. crisis had shown the company the challenges of transporting the RVM into new markets. It discovered that the business structures and power relations in the American market were too different for Tomra's idea of the RVM to fit in. Providing a high-tech machine was not enough to succeed in the American market. And the green wave by itself was not enough to generate sales. The machine had to be integrated into cultural, political, and organizational support structures, as a key component of a recycling system.

In the United States, Tomra wanted a larger part of these systems. In the second half of the 1980s, Tomra had consolidated its business model to ensure a stable economical basis. In the 1990s, however, the company started redefining its business model through a series of acquisitions of several material-handling companies. Neroc Inc. was the first of these, purchased in 1992. The company now saw system building as a way to reach a larger market: "Tomra's task is to help the world recycle. The natural consequence of this objective is to offer not just reverse vending machines, but totally integrated systems that make the entire recycling process more efficient."[108] Through these acquisitions, Tomra moved toward providing full container recycling value chain services. This included collection, pickup, processing, materials trading, recycling, and finally, production of new containers. Tomra also increased its control over the distribution network by purchasing the distribution companies. Tomra no longer sold its machines; it leased them for five years. Furthermore, the increased growth did not come from leasing the physical machines, but from "everything else": handling the returned empty cans and bottles, as well as settling the refund accounts through computers and other administration.[109] In 1996, 41 percent of Tomra's U.S. income came from materials handling and administration.

Tomra's efforts "on the other side of the wall" were particularly evident in Michigan, where it worked with bottlers to gain their support and cooperation in developing a "fully automated depot for cost-effective handling of empty beverage containers for the entire state."[110] Tomra designed the system to fully integrate everything from the grocery store reception of containers using RVMs to the back-end materials processing. Tomra thus aimed to become a total service provider of materials handling, accounting, and management of all types of beverage containers. It built a seventy-five-thousand-square-foot fully automated facility that could receive up to three billion beverage containers annually. To ensure a cooperative relationship with breweries and soft drink bottlers, Tomra negotiated with the Michigan Soft Drink Association and the Michigan Beer and Wine Wholesalers Association, covering most of the market. By creating such alliances, Tomra was able to construct a functional beverage container recycling system, which is now one of the most successful in the United States.

Greening the Green Machine

As we have seen, Tomra followed several strategies to directly reach the everyday environmentalist consumers who returned their beverage containers: new machine and store front designs, couponing, donation schemes, and integrated material management systems. Through the implementation of new physical and cultural elements in the RVMs, Tomra attempted to infuse the act of returning bottles with moral values.

Through the 1990s, Tomra reconfigured itself into an environmental company. The RVM was not only a recycling machine; Tomra now explicitly used it as a communication tool targeted at everyday environmentalists. Experiments with discount coupons, lottery tickets, and charity donations all show how Tomra actively attempted to influence users of the machine and build more extensive recycling systems. Tomra's management clearly thought of the company as an environmental company.

We have seen in this chapter how the company targeted culture and consumers—the intangible components of everyday environmentalism. The RVM became an increasingly multifaceted machine that could relate to many different identities and concerns: economy, the environment, ethics, and aesthetics. Indeed, Tomra attempted to tie all these sometimes conflicting discourses together in the machine. Tomra openly acknowledged that "our approach is to make recycling a way of life for the consumer."[111] In the next chapter, we will see how the company approached the more tangible aspects of everyday environmentalism through the building of infrastructure systems and new business and political alliances.

Chapter 7 Making Disposables Environmentally Friendly

IN THE SUMMER OF 2006, a Norwegian advertising campaign played on TV, in movie theaters, and online, featuring two men—Norway's "grand old man" Odd Børretzen and the young rapper Ravi—walking along a beach collecting empty aluminum cans.[1] Børretzen's warm voice intones, "One little discarded empty can like this one is nothing, you might think. It is like a drop in the ocean—but without drops in the ocean we wouldn't have an ocean at all—you haven't considered that, have you?" The two men get into a tiny red car and drive to a small traditional countryside grocery store. They then start singing a feel-good, summery song, with Børretzen counting and Ravi doing his distinctive rap:

"One—you get out the bottles and the cans."
They get the empty containers out of the car.

"Two—not just some, but they all go in the bag."
They put them in a plastic sack.

"Three—you find the deposit automat."
They walk through the little store, past the smiling cashier.

"Four—you take a little breather before the finale."
The camera shows a reverse vending machine, which is not identifiable as a specific brand.[2]

"Five—then you take the bottle or the can; you'll put it into the wall and then it will return."
Ravi then puts the bottles and cans into the RVM, and a sparkling new can magically appears in the cashier's hand.

Børretzen asks, "How does it feel, then?" and Ravi repeatedly replies, "It feels so good!" with a big smile.

Figure 13. Screen captures from the summer 2006 Resirk advertising campaign film, "It feels so good . . ." In the video, the RVM plays a central role in closing the loop between the consumer and recycled can. Courtesy of Norsk Resirk AS.

"Because a can that is returned enters into a cycle, just like a river that is useful perhaps forever—I wish it was me," says Børretzen.
They drive off through a typical Norwegian countryside landscape.

The deposit logo fades in: "Recycle everything—always!"[3]

This commercial was part of an information campaign sponsored by Norsk Resirk AS—a not-for-profit organization set up by Norwegian businesses to manage a deposit system for the recycling of disposable beverage containers: aluminum cans and disposable plastic. As we remember from the discussion of beverage cans in the 1970s in Chapter 2, a high tax kept disposable containers out of the Norwegian market because they were classified as environmentally unfriendly (and to protect the Norwegian beverage industry). What had happened that made beverage container recycling now "feel so good"? And why was it that the RVM played such a central role in the ad?

When some actors in the Norwegian beverage industry first attempted to sell beverages in disposable bottles in the late 1960s and early 1970s, the government mandated deposit-refund systems to prevent such a move. The industry was not content, however, with what it considered the inefficiency of reusable glass bottles. With changing market structures and an increased consumption of beer and soda, the costs and high energy use of transporting and cleaning heavy glass bottles soon became both an economic and an environmental problem. There is no doubt that aluminum cans and disposable plastic were economically acceptable—they were very practical and convenient. However, the challenge of the 1980s and the 1990s in the face of the "green wave" was to make them environmentally acceptable. In the case of aluminum, producers could highlight aluminum's recyclability and the energy savings from the material's light weight, which aligned the choice with the rising green wave of consumer environmentalism discussed in the previous chapter.[4] Industry interest groups needed to find an environmental solution for the disposables, to make them acceptable. To achieve this, they needed to consider the material, cultural, and political contexts of recycling on an equal basis. Either the containers had to fit within the established beverage container system for refillable containers, or a new, parallel system for disposable containers had to be implemented—without threatening the highly successful glass return system. Because public support for the existing glass system was strong, it was a significant challenge to remove the widespread association between disposables and littering.

A Case for Plastics

Some forms of litter are more disruptive than others, and plastic disposables first entered the Norwegian discussion as a way to reduce the amount of broken glass on streets and in nature. In the fall of 1972, Odd Schøyen Paulsen wrote an irate letter to the newspaper *VG*, complaining about the kids of today smashing glass bottles on the streets. Apart from highlighting the ever-present theme of young hooligans' not respecting their surroundings, Paulsen's letter brings up two noteworthy points. First, he argued that low deposit values were the main reason why such people smashed bottles instead of returning them. As we remember from Chapter 2, the industry-run glass bottle deposit system was at a critical stage at this time, with interest in cheap disposable bottles from breweries rising. If the breweries had been willing to set a higher deposit value on the bottles, the motivation to return empty bottles would have increased. Supposedly this would also work for young vandals. Second, and in a more realistic approach, Paulsen proposed that "if all the sugary drinks that the young drink came in plastic bottles, then at least we would not have to see all the broken glass shards on streets, roads, and in nature."[5] If you can't motivate hooligans to return their bottles, then at least you can frustrate their efforts to be a nuisance to others. In this case, the material qualities of plastic—in particular its resilience to breaking—came to represent a solution to bottle littering and vandalism. There was thus a clear undertone in Paulsen's letter: plastic bottles could keep in check some of the consequences of the lack of morality in today's society. The letter prompted a reply from an anonymous writer who stated that while this correspondent understood Paulsen's frustration over broken glass bottles, plastic bottles would not solve the littering problem: we would just be flooded with more plastic, which "we are blessed with more than enough of already."[6]

These early calls for plastic bottles tied into a strong public concern about littering in the early 1970s. Yet such concerns were initially not enough to introduce new material types into the Norwegian beverage container system. Given the strong preference for glass containers expressed by the labor movement, influential companies like Moss Glassverk, and the small bottlers and brewers of Norway, it is not surprising that plastic beverage containers did not enter into the discussion at the time. Plastic containers would have been one-way disposable containers rather than containers within the refillable-container system. As a result they did not serve the interests of any of the main actors in the discussion of the 1970s. The main proponents for disposable containers either aimed for disposable glass bottles or for disposable aluminum cans; plastic simply did not fit the existing system. Meanwhile, the refillable–glass bottle proponents acknowledged the problems caused by broken bottles, but chose to address this problem by mandating bottle deposits, as discussed in Chapter 2, which included raising the deposit from thirty to fifty øre for small bottles and from sixty øre to one krone for large bottles.

In short, disposable containers would not be seen as a viable alternative to glass bottles based solely on littering concerns. There also needed to be an economic incentive for the major businesses involved.

When plastic bottles again appeared in the discussion of beverage containers in Norway in the mid-1980s, the debate had little to do with littering and environmental concerns and all to do with market structures. While both grocers and consumers appreciated how light and durable these bottles were, particularly the large 1.5-liter bottles, grocers hoped that disposables could free them from their dependence on a small group of bottlers. The fee on disposable packaging favored the large, established breweries that had the expensive infrastructure for refillable-bottle handling in place.

In the grocers' view, there was simply too little competition in the beverage market. It was in their best interest to have a wide selection of goods to choose from at competitive prices. They argued that this was not the case in the beverage sector. For instance, the Norwegian Grocers' Association wrote a letter to the finance committee of the Storting, expressing its concern over the market dominance held by a few companies—most notably the Nora Group, which controlled about 50 percent of the Norwegian beverage market. The grocers accused the breweries of having divided the country geographically among themselves. Regional breweries in practice controlled the market in their own region through the economics of having to bring back empty bottles as required by the refillable–glass bottle system.[7]

As long as beverages came in glass bottles that had to be returned to the brewer for refilling—or a high tax would be levied—it was very hard for smaller soda producers to compete. The large, regional breweries had their own distribution systems, with trucks that delivered full bottles and picked up empty ones from grocery stores. Smaller companies had to rely on standard grocery distribution trucks, which were not able to bring empty bottles back.

In addition, acquiring the necessary amounts of bottles and crates to sell an independent soda brand represented a large investment. When the snacks company Maarud attempted to enter the soda market as a Pepsi franchise in 1981, they invested more than twenty million kroner in bottles and crates. In doing so, they went to war not only with Coca-Cola, but also with the Nora Group.[8] After only a year the war came to an end, and Maarud had to close its factory doors.[9] In what became a competition of consumer discounts and infrastructure expenses, a small company simply could not keep up with the giant company Nora, despite the strong support they had from grocers. Former Maarud CEO Bjørn Bunæs attempted to continue soda production under the name Maarud Mineralvann, but the new company was not able to come to an agreement with Pepsico International, which did not want to extend its loans to Maarud. Not long after, Nora purchased the Maarud bottling facility to produce Coca-Cola and its own brand, Solo.[10]

Bunæs did not give up that easily. He sent Norwegian ingredients to a bottler in England, who filled 1.5-liter plastic bottles with soda and then shipped them back to Norway.[11] The bottles could then be distributed along with regular groceries, without any need for returns. The disposable plastic bottles quickly grew popular, so Nora shortly began selling them as well. Because of the high environmental tax levied on these bottles (1.25 kroner for bottles up to a liter and 2.5 kroner for larger bottles), they ended up being quite expensive. Nora's productions costs were likely much lower than for Bunæs's imported bottles and Nora could afford to take a loss on the products in order to push competitors out of the market, thus Nora's price of twelve to thirteen kroner significantly undercut Bunæs's price of fifteen kroner.[12] The attempt to break the near monopoly of the large Norwegian bottlers thus failed, in large part because of the competitive constraints imposed by the beverage container recycling system.

The grocers did not initially argue for plastic bottles within the existing glass bottle return system; this would just play into the hands of the dominant actors in the beverage industry. Rather, they called for large, throwaway bottles that could be treated like regular household waste. Only in this way could they hope to challenge what they considered a "soda monopoly."[13]

INCLUDING PLASTICS IN THE EXISTING SYSTEM

Two separate proposals for the introduction of plastic bottles clashed in the mid-1980s. Both claimed that plastic bottles would be more environmentally friendly. However, one argued that refillable plastic bottles could replace refillable glass in the established deposit-refund system, whereas the other argued for the removal of the disposable-container tax and integration of disposable-container recycling rather than just refilling in the system.

Nora became a proponent of the first proposal. The company envisioned an expanded deposit system that included refillable plastic bottles instead of disposable ones. These refillable bottles were made of thick and durable polyethylene terephthalate plastic (PET), a thermoplastic polymer resin in the polyester family. To manage the repeated handling from washing and refilling, the plastic walls of refillable bottles are much thicker than those of disposable PET bottles. In addition to being lighter than glass and almost impossible to break, plastic bottles also do not generate as much noise as that of glass. This was a definite advantage in bottling facilities. Plastic suffers scuffs and scratches faster than glass bottles, which means that it is easy to identify bottles that have been in circulation for a while. After being refilled twenty-five times—the bottling facility marked the bottle with a small notch every time—the bottle was retired and recycled.[14]

While most plastics can theoretically be recycled, PET bottles are well suited for recycling, since they consist of only PET and can be easily identified and extracted from the waste stream. With PET bottles in a deposit return system, this becomes even easier, since the bottles are in their own waste stream, without any risk of contamination. This is important, because a single non-PET plastic container can ruin the recycling possibilities of a truckload of PET bottles.

While Norwegian grocery stores wanted disposable plastic bottles, Nora clearly considered imported plastic bottles undesirable competition. The company indicated to the Ministry of Finance that it would like to include refillable plastic bottles in the existing bottle return system. This basically meant that plastic bottles would be assimilated into the existing system that served the interests of the dominant industry actors.

A new RVM model provided the technological solution that made it possible to include the plastic bottle in the existing deposit system. Nora had cooperated with Tomra to develop a RVM that could handle both plastic bottles and metal cans. Since Nora was a major shareholder in Tomra, the two companies had many interests in common.[15] Tomra had started working on its new plastic-handling RVM in 1984, mostly targeting the market in the United States and other countries that were "trying to introduce one-way containers as recyclable on a voluntary basis." While doing so, Tomra argued for the benefits of "well-organized" deposit-refund systems, centered on the RVM.[16] The worst possible outcome for them was an increased number of disposable containers outside deposit systems, since these would never pass through the RVM.

Tomra displayed a prototype of the plastic bottle RVM at an American trade fair in October 1985.[17] It based this RVM, the Tomra 300, on the modular system it had developed for the Can-Can design.[18] The machine could handle both glass and plastic bottles, and became a lifesaver for Tomra in Europe. These new RVMs could use bar codes to identify bottles, which made it possible to accept bottles from only specific regions and certain producers—a necessity in the diverse European marketplace. Most important, it would not be possible to claim refunds for bottles sold before the introduction of deposits.

The other position advocated the inclusion of disposables in the system, primarily as a way to reduce litter. In 1985, Oslo city authorities asked the Ministry of Finance to consider removing the tax on disposable containers such as aluminum cans and plastic bottles.[19] Broken glass had become a major nuisance for the city authorities. Just as the grocers had done, the city pointed to the results from the Swedish aluminum can recycling system, where one had solved the problem "through using Norwegian Tomra's machines."[20] The relative success of Sweden's Returpack made disposable containers seem a viable environmentally friendly alternative to the existing refillable-glass system. From this perspective the glass system was not broken; rather, it was the broken glass that posed the main problem. Aluminum cans and disposable plastic bottles were unbreakable containers that could be recycled using an RVM-based system.

Even with the disposable container tax in place, consumers still generated disposables: twenty-five million aluminum cans were still sold in Norway in 1989, despite the high tax of 3.5 kroner, and by 1985 plastic bottles had achieved a 10 percent market share for soda despite an environmental tax of 1.25 kroner for small bottles and 2.5 kroner for large bottles. Many of these containers just "disappeared" after use, becoming litter. "It is easy to demonstrate that there are far more cans along the Norwegian wayside than in Sweden, where thirty times as many cans are sold," claimed Tore Planke in a newspaper interview, demonstrating how the visual impact of beverage container littering still carried weight in the discussion of beverage container recycling.[21] Grocer Trond Lykke—at the time chair of the packaging committee of the Norwegian Grocers' Association (Dagligvareforbundet)—called for an immediate expansion of the deposit system to include disposable containers. "What we want is the Swedish system."[22] In 1989, Lykke emphasized that while there were many reasons for why disposable packaging became increasingly popular in Europe, they did not want to move away from the existing system with glass containers. But to keep these disposable containers from despoiling nature, they had to be integrated into a deposit system.[23]

The two groups staked out their positions, and as we'll see, both ended up integrated into the existing glass container system model. But first we need to take a detour to Sweden, which yet again served as a model for integrating disposables into the RVM-based recycling systems.

Sweden Leads the Way

As discussed in Chapter 5, Sweden had implemented a deposit-return system for aluminum cans called Returpack in 1984. In 1988, Returpack ran another trial in Gotland, this time for PET plastic bottles. Again, it chose Gotland because it was geographically isolated: people did not bring bottles from the outside, yet it had high beverage consumption. Gotland had continually shown high recycling rates, indicating that recycling habits were clearly well integrated into everyday life. The results were good—about 70 percent of bottles were returned. It took six years, though, until the PET bottle deposit was introduced in all of Sweden, mainly because the different interest groups could not agree on a handling compensation for the grocers who had to receive the empty bottles. At the time, Sweden had several deposit systems running in parallel: one run by Returpack and others by individual bottlers and organizations, including Retur-PET, a system with four-kroner deposit for reusable PET bottles; Returplast, a system organized by smaller breweries, which had a fifty-öre deposit for reusable PET; and the Pripps system, with a fifty-öre deposit that could be claimed at only special Pripps brewery depots. Needless to say, the situation was messy for the grocers.[24] Until they could reach an agreement, Gotland kept the fifty-öre deposit. When the Swedish plastic deposit system called Returpack-PET finally started in 1994, it replaced the Returplast and Pripps systems, and introduced a deposit of one krone for small bottles and two kroner for large bottles.[25] Just as had happened when the aluminum can recycling system rolled out ten years earlier, there was some chaos when the new system began in 1994.

The relationship between the different types of disposable containers complicated the situation further. When Returpack was set up in 1984, the Swedish government and parliament had decided that Returpack was to give one öre per can to Keep Sweden Clean as a way to offset some of the littering that disposable aluminum cans would create. In July 1995, Returpack decided to refuse to do this any more.[26] They found that this donation gave them a competitive disadvantage compared with the rest of the Swedish packaging industry—an extra expense of nine million kroner annually. This made plastic bottles more attractive than aluminum cans. Returpack's vice president Rolf Andersson thus claimed that this extra fee was in conflict with EU regulations.[27] Needless to say, Keep Sweden Clean—which had no other sources of income—struggled financially and had to borrow money to keep afloat during the controversy. Returpack's move was unpopular, since Keep Sweden Clean organized high-visibility cleanup and awareness campaigns and environmental training for teachers. In the end, Keep Sweden Clean was strengthened by this crisis, since the Swedish parliament forced Returpack to continue contributing and the media attention led to new sponsors.

MAINTAINING SYSTEM BOUNDARIES

The introduction of beverage container deposits in a market means that the input and output to the return system needs to be strictly controlled. In 1995, Returpack decided to upgrade all RVMs with barcode readers in order to have this control. Swedish consumers had begun drinking so many foreign beverages (either imported after vacations or in nondeposit containers bought in stores) after joining the European Union that Returpack needed to be able to distinguish between Swedish and foreign containers in order to know which ones really had a refundable deposit. A 1996 estimate claimed that as much as 10 percent of all cans in Sweden had been purchased outside the Swedish deposit system. Returpack calculated that such containers had cost them tens of millions of kroner.[28]

Such a large order represented a major windfall for Tomra—the thirty-million kroner contract to upgrade fifteen hundred older machines contributed to a record result for the company in 1995.[29] In total, Sweden had four thousand RVMs, half of which were too old to upgrade. Tomra was in a position to deliver many of these RVMs as well, and saw increasing orders in Sweden.[30] The new RVMs not only had to be able to read the barcodes on the container labels, but also to connect to a central server to verify the deposit eligibility of the container. Returpack applied for financial support from the Swedish government to upgrade the system, but was denied.[31]

Although the barcode readers allowed Returpack to avoid paying refunds for foreign containers, the system still had to accept the foreign containers for recycling even though Returpack was not reimbursed for these expenses.[32] The idea was that the RVM needed to accept all containers in order to keep the system operational and encourage recycling.[33] Yet once the new RVMs started denying refunds on noneligible Swedish containers, consumers demonstrated what really mattered for recycling after the new system was in place: the returns of foreign cans dropped dramatically, even though consumption remained high. "The environmental friendliness of the Swedish people can be directly measured in the fact that they no longer get paid for foreign aluminum cans," a Swedish newspaper observed.[34]

Modifying the Norwegian System for Disposables

The plans to introduce disposable containers in Norway in the late 1980s caused a flurry of political, technological, environmental, and business activity. Many interests and agendas were involved, but as we will see, the environment was the rhetorical battlefield upon which they clashed. The existing deposit systems were reconfigured in the process, with the RVM firmly placed at the core of the systems.

A closer look at this discussion, the participants, and their motivations provides insight into the complex business-environment connection. The political parties, in particular the socialist parties, were split on the idea of a disposable-container recycling system. The involved organizations included aluminum producers, bottlers, labor unions, and environmentalist organizations. The discussion reveals that environmental arguments were not the only factors involved—industrial concerns and employment were just as important in decisions about establishing a disposables recycling system.

THE INTEREST GROUPS

The aluminum producers Norsk Hydro and Elkem joined together with other companies, including Tomra and PLM, to form a coalition to work on designing an aluminum can deposit-refund system in the fall of 1988. This was at the height of the "green wave" of consumer interest in environmentally friendly products. Hydro and Elkem wanted to change the public opinion of aluminum from being something that was environmentally destructive to being an environmentally friendly packaging. A recycling system for aluminum cans was an important element in this strategy. Such a program would also enable the companies to recover recyclable aluminum and thus save money.[35]

Aluminum producers had to argue that aluminum containers were not throwaways, but rather environmentally friendly, recyclable containers. "Aluminum is an environmentally friendly packaging, recyclable and economically sensible," argued Alexander H. Wirtz, director of Can Recycling Europe.[36] Wirtz said that glass bottles made economical and ecological sense only when sold within one hundred kilometers of the place of production. "The aluminum lobby must convince the politicians that recycling of aluminum cans is environmentally friendly," argued Wirtz.[37]

Grocers also found aluminum cans desirable. They were a lot easier to handle than glass bottles, both because they were lighter and because they could more easily be stacked. Independently of other initiatives, grocers had come together in their own committee to promote the use of aluminum cans, inspired by the Swedish Returpack system. In the spring of 1989, the Grocers' Association (Dagligvareforbundet) invited the Norwegian Association of Breweries and Mineral Water Producers (Norsk Bryggeri- og Mineralvannindustris Forening) to consider a Returpack-type system in Norway.[38]

When the grocers and the PLM-Hydro-Tomra group became aware of each other, they quickly joined forces in a consortium in June 1989.[39] This group produced a report on "deposit and return systems in the beverage sector in Norway" in September 1990.[40] This report recommended the establishment of a deposit system for one-way containers made of glass, PET, and aluminum. As part of the proposal, they wanted the 3.50 kroner tax on disposable beverage containers abolished. However, to remove an environmental tax supporting the successful system for reusable containers required the creation of an equally green system. The consortium thus proposed to found a neutral company called Resirk—an abbreviation of the Norwegian word for *recycle*—to administer and handle the logistics of the new recycling system. Resirk aimed to recover 90 percent of all disposable beverage containers sold on the Norwegian market.[41]

As in the 1970s, the beverage industry was not unanimously supportive of the consortium's plans. Tomra publicly attributed the resistance to the larger members of the beverage industry, in particular Ringnes, Norway's largest brewery.[42] Being such a large producer, Ringnes had a regional infrastructure in place that worked with returned glass bottles. Opening up the market for cans would only increase competition for them. The disposable container tax discouraged competition from international importers, who found it too expensive to transport glass bottles and the tax made it too costly to import drinks. Yet local industry interest in using them had been steadily growing. Many smaller domestic bottlers believed that using disposables would be a sound business decision because it was significantly cheaper to transport cans over the large distances that existed in Norway. This decreased transportation cost would open up the entire country's market to previously regionally restricted companies. For instance, the Hansa brewery in Bergen could transport canned beer and soda to the Oslo region, where most of the country's population live, for half the price they paid for transporting glass bottles.[43] The cost of canning machinery had also dropped since the 1970s, taking away some of the production advantage the large bottlers previously had.

Multinational bottlers like Coca-Cola and Pepsi supported Resirk because of the possibilities for centralization and economies of scale that disposables allowed for. While the international Coca-Cola Corporation has as its corporate policy to prefer voluntary consumer recycling to deposit systems, Coca-Cola's Norwegian president, Tore Bu, convinced the Atlanta management that Coca-Cola Norway should actively support Resirk.[44] Ringnes reluctantly supported the first Resirk application, supposedly after pressure from Coca-Cola Norway.[45] Pepsi also provided Resirk with some funding, but did not actively take part in the design of Resirk.

Complicating the picture further, labor interests divided the political parties' support for a disposables recycling system. The Norwegian Food and Allied Workers' Union (Norsk Nærings- og Nytelsesmiddelarbeiderforbund [NNN]) organized many workers in the beverage industry who would be affected by the Resirk system. NNN had a strong influence on the socialist parties. NNN used three main arguments against the Resirk plans. First, they argued that the Resirk system would threaten their workplaces. Between eight hundred and one thousand jobs would disappear when production shifted from reusable glass bottles to aluminum cans, and the new jobs created would be negligible. Second, the environmental consequences of disposable packaging would be worse than the old system. Finally, the proposed deposit system did not comply with the EC beverage container packaging regulations that specified that new systems that led to lower recycling rates in established systems could not be implemented. NNN also referred to Denmark, which had banned disposable containers.[46]

The NNN environmental arguments were based on a report that NNN commissioned on the Swedish experiences with aluminum cans to "evaluate the environmental consequences of introducing aluminum cans" to Norway.[47] The report concluded that consumer behavior differed with regard to the recycling of aluminum cans or bottles. Given a similar deposit, cans achieved a lower return rate than bottles. In Sweden, the return rate for cans in 1984–1986 was 65 percent, after the introduction of the Returpack system. When the deposit was raised from twenty-five to fifty øre in 1987, the return rate immediately increased to 75 percent, and to 82 percent in 1989. The recycling rates were higher in the larger cities than in the countryside.[48] This study's findings contradicted the extremely high 90–95 percent return rate that Resirk promised publicly. The report found convenience to be one of the most important factors for recycling. The survey mentioned RVMs, and stated that most consumers agreed that the RVM worked well.

Other labor groups supported the introduction of aluminum cans, in particular those who stood to gain from the production and recycling of cans in Norway. This split the organizations in the Norwegian Confederation of Trade Unions (Landsorganisasjonen i Norge [LO]).[49] PLM and Hydro Aluminium promised a can factory with 100–150 employees and a national recycling facility for five hundred million cans.[50] In addition, Resirk hired Gemini Consulting in 1992 to produce a report on employment consequences. Gemini concluded that 200 to 850 jobs might disappear, but that many new ones would also be created.[51] These statements clearly targeted the arguments of the union. Jostein Refsnes of Hydro argued that the beverage industry was already being restructured, and that any loss of employment should be attributed to these broader changes.[52] This was true, but the can discussion gave labor interests a focus for their resistance where they had solid arguments and a clearly defined opponent.

NNN worked hard to destabilize the environmental narrative that Resirk proposed by arguing for the uncertainty of the environmental consequences of a disposable-container recycling system. In November 1991, four thousand brewery workers went on strike for four hours to protest the proposed Resirk system, arguing that it would create environmental problems as well as threaten employment.[53] The Grocers' Association faced these arguments by claiming that Resirk would actually ensure a more decentralized brewery industry, and thus preserve employment in the districts. For this reason, Resirk had the small breweries on their side. Canning production lines were more affordable than in the early 1970s, making disposable containers more cost effective for them than reusable bottles. Seventeen small breweries argued in a letter to the Ministry of the Environment that they needed a Resirk system in order to survive.[54]

As we can see, it was a big challenge to create a new, untried system to compete with such a successful system as the reusable-bottle system. But industry structure and interests were changing, and new environmental concerns became relevant in the late 1980s.

What about the environmentalists, then? Like the industry groups, they were split on the issue. Several environmentalist organizations argued against Resirk on environmental grounds. The Norwegian Society for Conservation of Nature characterized the Resirk proposal as a logical industrial policy solution, but argued that it did not consider environmental issues.[55] Nature and Youth ridiculed Moss Glassverk for investing in facilities for the production of disposable glass containers: "Such a switch-over is pure suicide in a society where environmentalism is, and must be, growing."[56] Other organizations, like Bellona and the Green Warriors of Norway (Miljøvernforbundet), on the other hand, were positive about a Resirk system.

As we have seen, many parties had a stake in the question of whether or not to replace the high tax on disposable containers with a can recycling system in Norway. All sides used environmental arguments for their cause, but as we will see, labor interests came to dominate the discussion.

RESIRK'S LEGAL BATTLES

Establishing Resirk was a long, drawn-out process that would frustrate the members in the consortium immensely. Based on the recommendations from the 1990 report, Resirk sent an application to the Ministry of the Environment on November 30, 1990. The application asked for recognition of the industry-run Resirk system for recycling aluminum cans in lieu of the tax on disposable containers. The Resirk report had scheduled a start-up in early fall 1991, but the ministry spent more than a year debating the application and on December 13, 1991, they rejected it. While the ministry was positive toward the system, it feared that a deposit system for disposable containers might undermine the highly successful bottle system that was already in place.[57] Furthermore, it stated that there was too much uncertainty about the impact of the system on employment, as well as on Norway's relationship to the European Economic Community (EEC). Norway was currently negotiating with the EEC about a European Economic Area (EEA) agreement, which would include the management of packaging waste. The ministry thus advised Resirk to wait and try again later.

The rejection of the Resirk application clearly upset business interests, which now explicitly entered into political debates. Tomra's management slammed the government: "Gro Harlem Brundtland's government adopted the view that the uncertainty surrounding the consequences for jobs outweighed the benefits of an environmental measure which broke new ground."[58] By mentioning Brundtland, they wanted to remind the readers that as prime minister she did not live up to her status as former minister of the environment and global leader of sustainable development initiatives. Tomra had devoted "considerable efforts to promoting the introduction of a deposit on one-way beverage containers" and called the government's move surprising, "regrettable," showing "a lack of environmental concern," and "a large step backward" for "a nation committed to environmental issues."[59] In its choosing phrases like these, it was clear that Tomra did not hold back in its criticism of the government and that the company explicitly entered into political debates.

Despite the refusal, it seems that Resirk's application had some influence on the Ministry of the Environment. In a 1990 official Norwegian report on waste reduction and reclamation, published before Resirk's application, the ministry declared that the reuse of resources, including beverage containers, was preferred to recycling and energy reclamation.[60] This was in line with Norway's long tradition of reusing bottles. However, in a government's white paper, number 44 1991/92, published *after* Resirk had sent its application, the official stance on beverage containers had changed. Now, the main strategy was to "prevent generation of waste," and the tools for achieving this—"reuse, recycling, or reclamation"—were given an equal status.[61] The white paper mentioned deposit systems as an appropriate instrument for waste reduction, and that the government should encourage the private sector, that is, business, to maintain these systems. This change of heart gave Resirk some room in which to re-present their case.

CHANGING THE BASE TAX

On October 29, 1992, Resirk sent in its second application. The government replied that the Norwegian packaging tax system had to be changed before Resirk's proposal could be granted.[62] The Ministries of Finance and Environment would together consider a tax reduction for packaging handled in collection systems. As we remember from Chapter 2, disposable containers had been heavily taxed in Norway since 1970. As of 1987, this tax was 3.5 kroner.[63] However, the Resirk application suggested that there should be no tax on any packaging collected in its new system. Minister of the Environment Thorbjørn Berntsen argued for the removal of the tax, because "the can as an environmental problem has been considerably reduced, since it now can be recycled."[64] However, Berntsen had to avoid antagonizing labor interests (his primary constituency), and for that reason he eschewed taking a clear public stance in favor of Resirk.

To eliminate the 3.5 kroner tax on each can, the government instituted a fee system in which a tax was paid on the uncollected packaging. This differential tax, applicable to all packaging, was passed as part of the 1994 national budget. At a 95 percent recycling rate, the tax would completely disappear. The lower the return rate, the higher the tax. At 25 percent returns, the full tax of 3.5 kroner had to be paid. Along with this change, the Resirk application was approved on May 13, 1993. Newspapers called Tomra and Resirk the victors of the fight over cans. Tore Planke stated that the differential tax might turn into a Scandinavia-wide model for the handling of packaging that could influence EU recycling policy.[65]

However, the Resirk alliance celebrated too soon. Only six months later, the Storting added a seventy-øre base tax on all one-way containers. The base tax on one-way containers was a specific response by the finance committee and the majority of Storting to labor concerns, against the wishes of the Cabinet. In this way, reusable containers would have an advantage over disposable ones. The fact that the Storting at the same time asked the Cabinet to exempt Moss Glassverk from the base tax shows the strong influence of the glass industry.[66] When the base tax was introduced, it took the Resirk alliance by surprise.

The decision to institute a base tax was a hard blow for the Resirk consortium, which decided to dissolve the interim board.[67] It disagreed with the finance committee, which thought that a deposit system for disposable containers could be combined with the base tax. In a scathing press release, Resirk attacked this tax: "Encouraged by and in a close cooperation with Norwegian environmental authorities, business interests have for four years prepared the establishment of the Resirk consortium. The purpose was to collect and reclaim one-way packaging for sodas, squash, beer, and mineral water. A base tax will cause the Resirk system to not be economically viable." Resirk labeled the base tax "discriminatory," especially since both the Ministry of the Environment and the Norwegian Pollution Control Authority (Statens Forurensingstilsyn [SFT]) found one-way containers in a deposit system just as environmentally friendly as reusable containers. "The environmental engagement of Norwegian businesses would be stopped by a base tax that is not environmentally justifiable." Furthermore, Resirk claimed that the tax was in conflict with both the EEC waste directives and the EEA free trade rules.[68] The Norwegian Association of Food Retailers (Norges Dagligvarehandels Forbund) stated that the introduction of the base tax made it "difficult for businesses to take the authorities seriously when talking about environmental issues."[69] The Resirk coalition presented itself as an environmental innovation that could give Norway significant industrial competitive advantages.[70] In doing so Resirk wanted to emphasize that some parts of Norwegian industry stood to gain from this system.

The introduction of the base tax upset large parts of Norwegian industry, which attacked the Storting for not recognizing the consequences of its decision. Many companies canceled planned expansions related to aluminum and glass production and recycling, blaming the base tax.[71] These accusations, spearheaded by PLM's threat to cut 285 jobs at Moss Glassverk, were sufficient to make the government reconsider the base tax and the Resirk system within a year.[72]

In addition to the industry protests, for the next three years Norwegian authorities had to defend the decision to issue the base tax from the EEC. Resirk reported the decision to the EFTA Surveillance Authority (ESA), which concluded that the base tax indeed was in defiance of the EEA agreement. The government used environmental arguments to support the tax. ESA claimed that the base tax would target imported products almost exclusively and that there was no evidence that reusable glass bottles were more environmentally friendly than recycled aluminum cans.[73] ESA demanded that the Norwegian government remove all taxes on disposable containers.[74] Yet in 1997, the Ministry of Finance reissued the base tax as a policy instrument to support reusable beverage containers, arguing that Norway saw no reason to comply with a waste directive that was not even implemented in all EU countries. For instance, Denmark, Germany, and Finland were all exempted from the beverage container regulations in the EU waste directive. The ministry was responding to concerns that NNN continued to voice regarding the potential loss of jobs. NNN argued that the complaint to ESA was motivated by the strong economic interests behind Resirk.[75]

RESTARTING RESIRK

Encouraged by the ESA investigations and protests from Norwegian industry, the organizations behind the Resirk initiative decided to just go ahead and restore the interim board in December 1995, with the goal of starting up the recycling system. They realized that it would take several years for the political quagmire surrounding the base tax to be resolved. By starting the deposit system despite the base tax, Resirk hoped to prove that a recycling system for aluminum cans would work without weakening the existing system for reusable bottles. SFT approved the Resirk application on December 21, 1995, and made it clear that they expected a return rate of 90 percent.

Fifteen years after its original conception, Resirk finally launched.[76] On March 1, 1999, SFT approved a Resirk system proposal promising 90 percent return rates for aluminum and steel cans and 95 percent for PET bottles. Both NNN and Nature and Youth sent letters of complaint to the Ministry of the Environment, but without success.[77]

The general idea behind the system is that Resirk organizes one flow of money and one flow of materials. It was designed to overcome one of the specific shortcomings of the American deposit states that lacked a coordinating entity. Resirk works as a mediator between all the different actors. It is set up as a nonprofit organization, and it is designed to ensure that all actors in the system are interested in return rates that are as high as possible. Resirk is organized through memberships, which are open to all. Members get a reduction in the environmental beverage container tax, but have to pay an administrative fee per unit. All containers included in the Resirk deposit-refund system need to have an EAN bar code that has to be reported to Resirk before the container can be sold. The members pay the deposit with the administrative fee, and then the deposit follows the container through the entire system. In 1998, this administrative fee was twenty øre for an aluminum can or plastic bottle and thirty-two for a steel can. When the container finally ends up at the recycling depot, the deposit is refunded to the producer. Both the grocery stores and the recycling depot receive a small compensation for the extra work.

Interestingly, the grocers' handling compensation is higher if the store has a RVM. Resirk wanted to encourage a high rate of automation through this strategy: "Such an arrangement will reduce the risk for small stores and probably contribute to more RVMs and a higher acceptance of the Resirk system."[78] Through the online capabilities built into Tomra's machines, Resirk can continually monitor the container returns, which makes administration easy. Resirk also offered a financial subsidy to stores that wanted to invest in RVMs. RVM producers like Tomra thus stood to gain much from the tight integration of RVMs in the deposit system.

Resirk's income consists of the deposits, the administrative fee, and the scrap value of the containers. The expenses are the refunds, the handling fees/compensation to grocers and depots, as well as administrative costs. The administrative fee is adjusted so that the income and expenses are balanced. Resirk is intended to be self-financing and nonprofit.

Tomra's Involvement in Resirk

Where was Tomra in all this? The previous chapters have demonstrated how RVMs became part of recycling plans all over Europe. Along with the machine, businesses like Tomra gradually attempted to promote the political and cultural infrastructure that would encourage the use of these machines. The company kept a close eye on the world of politics, as it was here that the market potential for its machines was decided. Tomra's involvement in the Swedish Returpack system seems to have focused its awareness on these issues. Because Europe focused mainly on reusable bottles and the United States on recycling aluminum cans, Tomra consciously decided to develop technological systems that could handle both reuse and recycling. More important, the company targeted policy makers in order to encourage new deposit legislation.

Tomra had begun working with the aluminum industry early on. As we saw in Chapter 6, Sweden had come up with a system solution to the can problem in the combination of the Returpack organization and reverse vending technologies like the Can-Can. In France, Tomra cooperated with the aluminum industry in 1985 to place two Can-Can machines at the finish of each leg of the Tour de France to "demonstrate that aluminum cans are recyclable."[79] Tomra thus already had a track record of putting a positive spin on aluminum cans when the discussion of introducing a Norwegian disposable recycling system began. Through their involvement in the Swedish and American disposable recycling system, Tomra had developed a machine that handled aluminum cans and plastic bottles well. By encouraging the adoption of disposables, Tomra would be in a position to sell new or upgraded RVMs to all Norwegian grocery stores.

Tomra as a company newly focused on environmentalism actively promoted the aluminum can recycling system. Already in 1990, the Tomra Annual Report included a comparison of the old glass-recycling system and the proposed aluminum can system. It characterized the minimum 90 percent–return requirement for aluminum cans as a "good environmental alternative" to refillable glass. The current environmental tax on aluminum did not represent a "future-oriented solution to the problem of littering of nature."[80] By doing this, Tomra took a clear political stance on the issue of aluminum can recycling.

In spite of its early involvement, Tomra was not officially a member of the Resirk consortium. But when Resirk established several work groups to draft the 1990 proposal, Tore Planke chaired the technological committee. Although Tomra tried to keep a low profile in the Resirk discussions, newspaper articles connected it several times. For instance, in 1991 *Aftenposten* described Resirk as a coalition of Norwegian grocers and Tomra.[81] Tomra also contributed some financial support in the form of paying for a secretary (recruited from Tomra) to help with the final application in 1996–1997. Tomra gave the third-largest individual contribution, and approximately 10 percent of the total sum.[82] By supporting the consortium, Tomra was helping build a coalition that would lead to increased sales of its machine.

Tore Planke was the primary designer of the model underlying the Resirk system.[83] The 1986 crisis had given Tore a new role within the company. While Petter Planke decided to leave Tomra, Tore remained as technical director and became more active as a lobbyist and networker, roles he had increasingly filled during the 1980s. Like many of the engineers whom historian of technology Thomas Hughes describes, Tore became more of a systems builder than an inventor.[84] In 1994, he argued that the Resirk plans failed because of a lack of a good lobbyist and a united front among the grocers. None was willing to be Resirk's public spokesperson.[85] For this reason, Tore came to represent Resirk in public discussions, flanked by Resirk secretary Erik Røsrud.

By mobilizing his knowledge of the integration of technological systems and the environment, Tore was able to position himself as an expert. Tools like life cycle analyses and graphs demonstrating the relationship between deposit values and recycling rates established connections between Tomra's technologies and the environment. By presenting a certain interpretation of the world, in which the RVM could transform disposable containers into environmentally friendly products, Tore positioned Tomra's products at the center of the Resirk system. This was precisely what he had not managed to accomplish in New York ten years earlier. Now, Tomra built systems rather than individual machines.

The close connections between Tomra and Resirk had the potential of becoming a liability. Tore Planke's extensive technological and political knowledge was extremely valuable for Resirk in the start-up phase, but it became increasingly important for Resirk to demonstrate its independence from Tomra.

In the early Resirk discussion, Tomra CEO Svein Jacobsen had taken a clear stance. For instance, the 1991 annual report clearly stated that "during the year Tomra devoted considerable efforts to promoting the introduction of a deposit on one-way beverage containers in Norway."[86] Furthermore, Tomra's newsletter, *Tomra News*, acknowledged in 1993 that the company had "a clear opinion for or against can deposits in Norway."[87] The problem was that since Tomra as a company stood to gain so much from a Resirk system, it did not have much credibility when arguing for the environmental benefits of the system. Instead, the more Tomra argued for the system, the more other actors resisted.

Tomra's top leadership soon chose a more neutral path. Erik Thorsen, who had been Tomra's chief financial officer since 1990, replaced Svein Jacobsen as CEO in April 1996.[88] During Thorsen's tenure as CEO, Tomra's annual reports downplayed the company's political role. In 1997, the company stated that it had "no ambitions to lead or influence policy makers about how to handle recyclable packaging. That's beyond the mandate of a commercial, worldwide enterprise, nor would it be an appropriate way of spending the money of our shareholders. We do, however, offer practical and cost-effective solutions that are readily adaptable to local ordinances and market-driven needs."[89] This was a dramatic shift in the official line. Through the sudden change in corporate policy, Tomra hoped to avoid antagonizing its customers and shareholders. Resirk was hardly mentioned again in the annual reports.

When Resirk finally started up in 1999, Tomra publications merely acknowledged the fact that the system was operational and that the company foresaw a large growth in Norway because of it. Jarle Grytli, who was then CEO of Norsk Resirk (and a former Tomra employee), was granted one page in a 1999 newsletter to describe Resirk. He does not mention the process in which Resirk was created, other than state that it was "almost ten years of governmental and business political struggle."[90]

In spite of trying to leave the politics behind, Tomra did have business to gain from Resirk through the RVM. Resirk wanted—and needed—to encourage competition when Norwegian grocers started buying new RVMs to comply with the Resirk system. A plethora of small companies entered into the fray: Swedish Nimo-Verken, Microlux AB, and Canmatic had all been encouraged by Returpack to develop affordable RVMs that could handle disposable containers, so they had machines available; the U.S.-based Envipco wanted to win a share of the new Norwegian market; and the small Norwegian company Repant launched its own machine. Resirk aimed for a cooperative relationship with these companies, but required a minimum functionality of the machines.[91] Tomra positioned itself as "the only supplier with a machine that can handle all types of containers in the market"—the Tomra 600.

The T600 launched on April 1, 1997, on Tomra's twenty-fifth anniversary. It was developed to handle aluminum cans and plastic disposable bottles in Norway and was the most technically advanced machine Tomra had ever made. All beverage containers could be inserted horizontally through a round hole in the wall, solving the stability problems that arose in the older machines where bottles had to be vertical when inserted. A high-resolution charge-coupled device (CCD) video camera paired with computer processors scanned the inserted beverage containers at high speed and could identify all sorts of containers accurately.

Online functionality was key to this model change. It permitted programming, monitoring, fault diagnosis, software upgrades, and data collection through a modem to a central computer. These online systems were developed for the American market as part of Tomra's move toward becoming a total system provider. In 1995 and 1996, Tomra upgraded fifteen hundred Swedish machines for Returpack to include online functionality. The Norwegian Wine Monopoly also ordered fifty-seven machines with online functionality, the first order in Norway. In 1995 Tomra upgraded its software, which could be retrofitted on old machines to add online functionality. The online capability was vital to the Resirk model because it allowed real-time monitoring and administration of container returns.

Embedding the Environmental Machine in Sociotechnical Systems

This chapter has built upon the analysis of business environmentalism in Chapter 6, but seen from another side. Tomra's involvement in the Resirk discussion is not about the ideological greening, but about making systems that work. However, to get to this point, Tomra needed to do work on both an ideological and a technical level. Aluminum cans had to become technically manageable and socially acceptable.

Jørgen Randers stated in 1997 that Tomra had to "force good solutions on the market."[92] However, even though Tomra had become the largest and most successful RVM manufacturer in the world, it did not have enough power to force anything on the market. Instead, they had to negotiate and build networks of actors and interests. Only in this way could the RVM become the obligatory point of passage for disposable containers.

It is important to stress that Tore Planke was not the only person behind the Resirk system. Neither was Tomra the only one who benefited financially from it. Yet Tore Planke's influence gave the system its final shape. Tomra interacted with environmental policy makers in Norway to create a disposable-container recycling system that placed technology in a central role. This chapter has shown us that the manufacture of technology can have just as much impact on environmental policy as direct, political work. By providing technologies for implementing the bottle bills, Tomra influenced the form and content of such policies. From working in the backroom of grocery stores, Tomra and Tore Planke had to become politicians, working to encourage the creation of the Norwegian disposable-container recycling system.

While the Resirk system on the surface seems to be set up to control and regulate a flow of materials, a closer analysis of the system demonstrates the complex, hybrid nature of modern environmental policy. Like everyday environmentalist actions, environmental policy is not "pure." Instead, it gains its strength from this blend of material, cultural, political, and business elements. Technological solutions, in this case in the form of an RVM, were integrated into environmental policy in order to achieve political and business goals. Since 1999, the Resirk system has become ingrained in Norwegian culture. Disposable beverage containers now feel "so good" for the everyday environmentalist. They are both convenient and environmentally friendly—but only as long as they are returned to the store in an RVM. The machine thus absolves consumers of any environmental sins. As the 2006 video advertising campaign demonstrated, the system allows the everyday environmentalist action of returning cans in order to recycle the material into a new, shining one. And there, as a key actor in the system, is the RVM: a green machine.

Chapter 8 Message in a Bottle

Climate change looms large in the current public discourse on environmentalism. During the Easter holidays of 2007, Resirk ran their new information campaign, stating how "Earth has a fever. We must all contribute to reducing energy consumption and emissions. Returning your beverage containers is a small but important contribution to a big issue."[1] The everyday environmentalist ethos could not be stated any clearer: consumers could now combat global warming by recycling their empty bottles and cans. In the broader discourse on global warming, consumers are encouraged to use low-energy lightbulbs, drive hybrid cars, and buy carbon credits to offset vacation flights. Rather than think of these initiatives as charades, we should try to think about the role technology can play in connecting these small everyday actions to large-scale environmental benefits. Since our consumer lifestyles are contributing to environmental damage worldwide, I believe that realistic environmental measures must embrace—rather than deny—consumerism. I do not believe that people will stop consuming bottled beverages, so if we want to reduce the environmental footprint of that activity, we need to control the packages and the way they are handled. As companies and consumers strive to become carbon neutral, the lessons of everyday environmentalism and tools like the RVM are more relevant than ever.

We have seen throughout this book how the RVM was much more than a hole in the wall as it moved through material, political, and cultural spaces. In the material space, the RVM was a physical artifact doing a physical job of handling empty bottles. In the political space, it was a tool for political actors to achieve political aims, including economic and environmental goals. The cultural space is where the RVM brought together the ideologies and practicalities of recycling. The RVM's movements between these spaces were foreseen only partly in the early 1970s. Yet the foundation had been laid with the first prototype Tomra RVM. The RVM evolved in an intricately interwoven relationship with developments in a larger system of beverage container technologies; the legislation instituted to manage these containers; the companies and organizations involved in producing, distributing, and selling beverage containers; and finally the ways in which consumers consumed and disposed of their empty bottles and cans.

The Many Faces of the Machine

I have followed the reverse vending machine and its creators in their research and development laboratories, through society, into policy, across national borders, and into consumers' lives. We have seen throughout this book how various actors have entered and left the story, having shaped and being shaped by the RVM. Tomra latched onto ideas about littering in the 1970s, the landfill crisis in the 1980s, and global resource use in the 1990s to help redefine the bottle problem and its solution to it. All along, while the technological platform has become more advanced, what the machine does *functionally* is to receive bottles and give back a receipt in order to support the recycling system. In this sense, I have told not as much a history of technological change as a history of the changing cultural identities of technology and the rise of consumer environmentalism.

The problem of bottles appears in multiple discourses and contexts that are gradually interwoven throughout this book.[2] We saw in Chapter 2 how empty bottles were simultaneously framed as a visible environmental problem, an infrastructure problem, an industrial opportunity, and an object for taxation in the Norwegian context. When Tomra began producing its RVM, the Planke brothers mobilized resources from only one of these discourses—empty bottles as an infrastructure problem. This allowed them to target grocers and their problem with high precision. While the Planke brothers thought of the RVM as a machine solving environmental problems when judging the market for the machine, they did not market the machine as such. The green of environmentalism in the early 1970s could not be exchanged for the green of money.

When Tomra began exporting its machine beyond the borders of Norway, the RVM entered into new contexts and cultures of recycling. Tomra discovered how the problem of bottles was not only widespread, but also more complex than its original machine could handle. However, the business opportunities were too promising for the ambitious company to pass up. So Tomra developed the self-learning RVM, which was flexible enough to manage this new world of bottles. Tomra packaged the RVM as a flexible, high-tech machine that could solve the problems of grocers worldwide.

In the 1980s, the RVM moved into a larger discourse on industry building and high finance. Through Tomra's European successes, the RVM became a money machine that promised to make investors rich. Ambitions for growth moved to the top of Tomra's corporate agenda and have remained there. Tomra assumed that the machine's ability to handle all types of bottles would allow it to fit into all kinds of cultures. Its experience in the American market proved it wrong. The Scandinavian culture of recycling had become intertwined with the RVM, but inserting the RVM into a different culture of recycling did not necessarily produce the symbiotic relationship that would allow the machine to succeed.

A new generation of investors and professional managers rebuilt Tomra after the disaster in the United States by framing the RVM within the growing discourse on sustainable development. Beginning in the late 1980s, empty beverage containers became connected to the global environment through discussions of sustainable development and the interconnectedness of local landscapes and the global environment. Tomra's corporate publications from the 1980s and 1990s repeatedly illustrate this idea. In their so doing, Tomra actively promoted its technology as a way to perceive, order, manage, and navigate through the recycling dilemmas, for both customers and policy makers. As we saw in the development of the Resirk system, technology became a facilitator for the realization of environmental policies, and as such, the enabler for everyday environmentalism. The RVM became a green machine that could help offset the environmental costs of rising beverage consumption in modern consumer society. The cultural identity of the machine and the definitions of what constituted environmentally friendly behavior developed in a mutually dependent relationship.

As the connections between beverage containers and the environment shifted, the RVM had to be repositioned. Throughout the RVM's history, its success depended on interactions between the machine and the market. The "best available technology" that Tore Planke wanted to use had to be translated into products that covered grocers' needs in 1972. Environmentalism had to be translated into increased sales in the late 1980s. The RVM had to be translated into a tool for environmental policy in the Resirk discussion of the 1990s.

By weaving in and out of different discourses, the RVM went from being an efficient infrastructure machine to being a green and efficient infrastructure machine, from a local to a global scale.[3] In a way, it is possible to say that the RVM was always an environmental machine, but that society was not ready for the green RVM when Tomra started. Tore and Petter Planke wanted to change the world, but they had to wait for the world to change first.

Anonymous Automats and Visible Visionaries

The RVM is an anonymous technology, which are "humble things, things not usually granted earnest consideration, or at least not valued for their historical import," as Siegfried Gideon described them. In Gideon's view, these humble objects in the aggregate "have shaken our mode of living to its very roots."[4] Unlike the car, the RVM has not radically transformed the human-built world. Unlike the telephone, it has not changed the way humans interact and communicate with each other. Unlike the Internet, it has not altered the way information spread throughout the world. The RVM's influence is more subtle and diffuses through the pervasive anonymity of the machine (and the system in which it is embedded) and the visibility of the visionaries who created it.

The RVM has become an obligatory point of passage for container recycling in many markets, yet an almost invisible one. Bruno Latour uses the term "obligatory point of passage" to discuss how the meaning of a scientific fact becomes accepted so that anyone who wants to talk about a certain topic needs to include this fact.[5] In countries and states that have a deposit-refund system for beverage containers in place, the RVM has become a obligatory physical point of passage. Billions of empty beverage containers pass through the RVM to be reused or recycled. After the green turn in the 1980s, the RVM also became an obligatory passage point through which aluminum cans must pass in order to become green, to be under control, something clean and no longer dangerous. As litter, cans were matter out of place. When sent through the machine, they were no longer a problem.

There is power in becoming a point of passage as well. The RVM helps redefine what actions are environmentally correct; bottles and cans returned for a refund benefit the environment and those thrown away do not. The machine has the power to translate good environmentalist intentions into physical reality through economically viable arrangements. As such, it represents a model for how policy makers, NGOs, and businesses can think about environmental challenges and goals. The RVM is the glue that keeps the entire beverage-container recycling infrastructure together. By reducing complexity and providing convenience for its users, it can enroll and aggregate their actions into a larger system. At the same time, by giving its users access to this technological system, it helps transform empty-bottle returns from individual, largely symbolical actions to effective environmental measures.

While the machines have become largely invisible, their creators have found themselves in the limelight. The Planke brothers had to champion both the machine and cultural structures that would support it, so they actively cultivated a positive relationship with the Norwegian media. The story of a small Norwegian firm making it big in the world market struck a chord with the Norwegian public. Newspapers and magazines regularly interviewed the Plankes, and Petter found himself a leading spokesperson for the Norwegian high-tech industry.

The Planke brothers appear as central characters in my narrative because I follow the RVM, and they refuse to let go of the machine—they want to control what happens with it. Yet the Planke brothers could not become powerful enough to determine the entire system. They could not lay the entire system under their control, like Ford or some of Hughes's system builders. Instead, they constructed the machine to align with the interests of other actors in the system. It made it possible for these other actors to achieve their goals, to make some solutions seem more possible, to make some actions seem more convenient. The Plankes could at times slightly redirect the system, which requires sensitivity and expertise. The discussions of the Returpack and Resirk systems show how they were an act of system building with many actors. There was no master builder in the center, but rather a distributed effort attempting to align, recruit, and mobilize the interests and resistance of many groups. While I have told this story mostly from Tomra's point of view, equally valid stories can be told with other actors positioned at the center.

As we have seen, successful recycling systems must enroll and mobilize a broad network of actors. I have attempted to see the actors in this story as master marketers as much as seeing them as master builders. The story has touched on the recruiting and mobilizing of users, and how people tried to keep their support. This book has allowed us to see which resources and interests the machine's producers mobilize, as well as others who mobilized the RVM.

Throughout the RVM's history, different actors have delegated tasks to it. We saw that the first important group were grocers, who needed an infrastructure to handle returned bottles and cans. When the Plankes designed the RVM in the early 1970s, the purpose was to help grocers do manual work in the backroom. Bottles made for messy and demanding work; replacing humans was cost effective and also more convenient for both employees and customers. The needs of grocers was addressed by making consumer-users return their bottles correctly and conveniently. Tomra and other RVM producers taught consumers to recycle through the machine interface and the media. Later, recycling advocates and policy makers promoted the implementation of RVMs in the political arena. Policy makers created legislation that supported or required the use of the RVM. This diverse group of users and functions turned the RVM into a multifaceted machine, but also created tensions between the different functions.

Lessons from the Intersection of Histories

By looking at the history of the RVM and its place in sociotechnical systems, we have seen that everyday environmentalism works by making environmentally friendly actions convenient, profitable, and pervasive. This has broader implications for historians looking at business, environment, and technology history, as well as for policy makers in the twenty-first century.

First, this book has demonstrated the integrated nature of political goals and business interests. One cannot succeed without the other. There are difficulties with transporting technological products and business models between different markets. Tomra could not simply transport its RVM without considering the systems and social networks into which it is placed. In the New York case, Tomra attempted to insert its technology into a noncooperative environment. In the Norwegian case, the system was carefully designed around the machine instead. In both cases, the outcome of environmental policy initiatives depended on the strength of business alliances. To be successful at selling its technological products, Tomra had to become engaged in political work. By mobilizing a wide range of resources and allies, the company managed to maneuver through and stabilize a complex network of social, cultural, economical, political, technical, and environmental factors. Rather than wielding power openly, as, for instance, Coca-Cola did in New York, Tomra had to negotiate, nudge, suggest, and leverage every resource it could muster. In a lecture Tore Planke gave in 1995, he listed three objectives for success: cooperation, cooperation, and cooperation.[6] This was clearly something he had learned from the U.S. failure, and that he had worked hard to ensure in designing the Norwegian system. Tomra's success in one deposit market, but not in the other, demonstrates that cooperative business alliances are critical to creating the infrastructure that makes environmental policy work. Business historians need to recognize the role businesses have in setting policy goals and implementing them, and that this is an important subject of inquiry.

Second, the case of Tomra demonstrates that business can play a role as environmentalists, not just as ravishers of pristine nature. Businesses align themselves with environmental goals rarely as a way to be "good" (even though it might certainly inspire entrepreneurs), but do so because they see it as offering a competitive advantage and a market opportunity. This strategy complicates business and environmental ideals and requires ideological work—it is not enough to have good products and to be cheap—companies must also maintain a solid environmental record. When environmental historians have studied environmentalism, there is a tendency to judge environmental initiatives by the means and not the ends. They often question the motives of businesses that will make money from environmental actions. Should this be the case? Perhaps the Tomra example shows that businesses can make money and do something positive for the environment at the same time. Everyday environmentalism is rooted in the interplay of consumer lifestyles and business opportunities.

Third, the history of the RVM has demonstrated how broad ideological shifts have a direct effect on sociotechnical systems. Historians of technology should look at not only how designers make machines and what users do with a machine when they get one, but also at why they make the decisions. The larger changes in environmental thinking had a direct bearing on both the design of the RVM and its cultural framing. Bernie Carlson has argued that Thomas Edison's movie production efforts failed because Edison's frame of meaning for movies did not align with larger cultural shifts; people wanted romance and action, while Edison wanted education.[7] The Tomra story shows how both the company's frame of meaning and the social one changed over time and how technologies must adapt to these shifts in order to be viable. Historians of technology should look at environmentalism not as a stagnant ideology but rather as a dynamic and changing basis for technological change. Technology that appears to be "environmental" today might not always have been so framed.

The case of Tomra and the RVM moves beyond contributing to historical discourse; it should contribute to current approaches to environmental policy. This book has shown how environmental thoughts can be converted into environmental actions. The good intentions of everyday environmentalist actions cannot be separated from the technological, political, and business structures that support them. At the time of Tomra's founding, a deposit system for glass bottles was in place. The Planke brothers did not have to think about the political systems that their machine became a part of. With the introduction of aluminum cans and disposable plastic bottles, the situation was different; disposables had to become environmentally friendly. The RVM could transform the throwaway containers into green ones, but only when they were integrated into a functional political system supporting recycling efforts. In these markets, the RVM became a tool to translate environmental policy into consumer actions.

Bottle bills are still hotly debated all over the world. Several American states are considering introducing or modernizing some states' bottle bills to include new container types, higher deposits, and revamped schemes for handling unclaimed deposits. While some states have actually succeeded in expanding bottle bills, such as the New York Bigger Better Bottle Bill from 2009, it is also clear that resistance to mandatory bottle deposits is as high as ever. Yet the proponents are gaining ground—a showdown is imminent in many states, among them Tennessee, Minnesota, and Massachusetts. In France, an unsuccessful proposition was presented in 2007 to the French National Assembly to require a deposit on all glass bottles of beer—returning to a deposit system that France had abolished in 1989.[8]

A 2007 *New York Times* article mentions that the only well-functioning American deposit system can be found in Michigan. While they attribute Michigan's success to the higher deposit value of ten cents rather than five cents, I would argue that it is significant that Tomra runs the system and has designed it to align with business interests and RVM technology.[9] One of the lessons from this book that bottle bill activists need to take to heart is that beverage container legislation is not enough to bring about recycled bottles; it requires a system built on active support from the key business actors and convenience for consumers. If you don't get the container back from the consumer, the loop is broken; and when the consumer returns his or her bottle or can, the system needs to be designed to accept it.

Environmentalist philosophies like Arne Næss's deep ecology ultimately have their roots in utopianism. They require people to make sacrifices—to dramatically change their lives and to challenge powerful vested interests in business and politics. I do not believe we will be able to achieve such utopias on a large enough scale to be effective on this side of enlightened world dictatorship. Everyday environmentalism, on the other hand, has the potential to spread wide and shallow. But it has to be convenient. Policy makers need to require the business world to be environmental; at the same time, it must be up to business how to achieve these goals. One of the greatest contributions that businesses can offer is support for the technological infrastructures that make environmentalist actions feasible and convenient. By aligning businesses, policy makers, and consumers, we can create the sociotechnical infrastructures that encourage and support everyday environmentalism.

Notes

Chapter 1

1. Frank Ackerman, *Why Do We Recycle? Markets, Values, and Public Policy* (Washington, DC: Island Press, 1997), 123.
2. "Recyclers Scrapped from Beijing ahead of Olympics," *International Herald Tribune*, July 9, 2008.
3. However, the bottled water industry sued the state of New York in May 2009, demanding a postponement of the new law.
4. Container Recycling Institute, "Recycling Rates by Materials and Class, 2006," http://www.container-recycling.org/facts/all/data/recrates-depnon-3mats.htm.
5. In 2009, 396 million single-use aluminum cans and PET bottles were recycled in Norway. Resirk, "Fakta og tall," http://resirk.no/Fakta-og-tall-52.aspx. In addition, reusable glass and PET bottles still have a significant share of the Norwegian beverage container market, although the total number is unavailable.
6. "Stadig en skrue løs," *Dagbladet*, December 31, 2004.
7. The Norwegian Climate and Pollution Agency, "Gjenvinner flere flasker, kartonger og bokser," July 6, 2009, http://www.klif.no/no/Aktuelt/Nyheter/2009/Juli-2009/Gjenvinner-flere-flasker-kartonger-og-bokser/.
8. Melvin Kranzberg, "Technology and History: 'Kranzberg's Laws,'" *Technology and Culture* 27 (1986): 544–560.
9. See Dan Baum, *Citizen Coors: An American Dynasty* (New York: William Morrow, 2002); Tachi Kiuchi and William K. Shireman, *What We Learned in the Rainforest: Business Lessons from Nature* (San Francisco: Berrett-Koehler, 2002).
10. See, for instance, John Tierney, "Recycling Is Garbage," *New York Times*, June 30, 1996.
11. See, for instance, Ackerman, *Why Do We Recycle?*
12. The discussion on "the water wars" in the developing world is one example: Vandana Shiva, *Water Wars: Privatization, Pollution, and Profit* (London: Pluto Press, 2002).
13. Thomas P. Hughes, *Networks of Power: Electrification in Western Society, 1880–1930* (Baltimore: Johns Hopkins University Press, 1983); Thomas P. Hughes, *American Genesis: A Century of Invention and Technological Enthusiasm, 1870–1970* (New York: Viking, 1989).
14. Katie Kelly, *Garbage: The History and Future of Garbage in America* (New York: Saturday Review Press, 1973); Louis Blumberg and Robert Gottlieb, *War on Waste: Can America Win Its Battle with Garbage?* (Washington, DC: Island Press, 1989); Jennifer Seymour Whitaker, *Salvaging the Land of Plenty: Garbage and the American Dream* (New York: William Morrow, 1994); Elizabeth Royte, *Garbage Land: On the Secret Trail of Trash* (Boston: Little, Brown, 2005); Elizabeth Grossman, *High Tech Trash: Digital Devices, Hidden Toxics, and Human Health*

(Washington, DC: Island Press, 2006); and Heather Rogers, *Gone Tomorrow: The Hidden Life of Garbage* (New York: New Press, 2006).

15. Martin V. Melosi, *Garbage in the Cities: Refuse, Reform, and the Environment: 1880–1980* (College Station: Texas A&M University Press, 1981); Matthew Gandy, *Recycling and the Politics of Urban Waste* (New York: St. Martin's Press, 1997); Martin V. Melosi, *The Sanitary City: Urban Infrastructure in America from Colonial Times to the Present* (Baltimore: Johns Hopkins University Press, 2000); Benjamin Miller, *Fat of the Land: Garbage in New York the Last 200 Years* (New York: Four Walls Eight Windows, 2000); and Carl A. Zimring, *Cash for Your Trash: Scrap Recycling in America* (New Brunswick: Rutgers University Press, 2005).
16. Andrew Szasz, *EcoPopulism: Toxic Waste and the Movement for Environmental Justice* (Minneapolis: University of Minnesota Press, 1994); Susan Strasser, *Waste and Want: A Social History of Trash* (New York: Henry Holt, 1999); William Rathje and Cullen Murphy, *Rubbish! The Archeology of Garbage* (Tucson: University of Arizona Press, 2001).
17. Michael Bess, *The Light-Green Society: Ecology and Technological Modernity in France, 1960–2000* (Chicago: University of Chicago Press, 2003), 4.
18. Deborah Lynn Guber, *The Grassroots of a Green Revolution: Polling America on the Environment* (Cambridge: MIT Press, 2003), 177.
19. Hal K. Rothman, *The Greening of a Nation? Environmentalism in the United States since 1945* (Fort Worth, Tex.: Harcourt Brace College, 1998), 5.
20. "Gore's Bold, Unrealistic Plan to Save the Planet," *Time*, July 18, 2008.
21. Hughes, *Networks of Power*; Hughes, *American Genesis*; and Thomas P. Hughes, *The Human-Built World: How to Think about Technology and Culture* (Chicago: University of Chicago Press, 2004). See also Pär Blomkvist and Arne Kaijser, *Den konstruerade världen: Tekniska system i historisk perspektiv* (Stockholm: Brutus Östlings Bokförlag Symposion, 1998).
22. Of course, the technological challenges of developing fuel cells or batteries with sufficient capacity should not be understated.

Chapter 2

1. The øre is the smallest Norwegian monetary unit. There are one hundred øre to one krone.
2. Byhring repeated his performance in 1971 at the Schou Brewery's 150th anniversary. Quoted in Øystein Øystå, *Brygg, brus og bruduljer: Bryggeri- og mineralvannbransjen i Norge 100 år* (Oslo: Bryggeri- og mineralvannforeningen, 2001), 152.
3. Christine Myrvang, Sissel Myklebust, and Brita Brenna, *Temmet eller uhemmet: Historiske perspektiver på konsum, kultur og dannelse* (Oslo: Pax Forlag, 2004).
4. Prior to 1960, the Norwegian government required a license for all cars paid for in dollars, i.e., all cars except those produced in Norway (few) and Russia (low status); most of these licenses went to doctors, traveling salespeople, and others with a documented business need for cars: Per Østby, "Flukten fra Detroit: Bilens integrasjon i det norske samfunnet" (PhD diss., University of Trondheim, 1995).
5. Norway, Ministry of the Environment, *Resirkulering og avfallsbehandling II*, NOU 1975:52, 119.
6. Mark Pendergrast, *For God, Country, and Coca-Cola: The Unauthorized History of the Great American Soft Drink and the Company That Makes It* (New York: Basic Books, 1993), 295.
7. Alan MacFarlane and Gerry Martin, *Glass: A World History* (Chicago: University of Chicago Press, 2002), 10.

8. William S. Ellis, *Glass: From the First Mirror to Fiber Optics, The Story of the Substance That Changed the World* (New York: Avon Books, 1998), 38.
9. Beverage World, *100 Year History of the Beverage Marketplace, 1882–1982 and Future Probe* (New York: Beverage World, 1982).
10. Resirk CEO Jarle Grytli, interviewed by author, January 16, 2006, Oslo, Norway, digital recording.
11. See Langdon Winner's classic argument on how some technologies encourage or require certain ways of organizing society in "Do Artifacts Have Politics?" in *The Whale and the Reactor: A Search for Limits in an Age of High Technology* (Chicago: University of Chicago Press, 1986), 19–39.
12. E. Petersen and C. J. Arnholm, *Frydenlund Bryggeri: 100 år 1859–1959* (Oslo: Frydenlund Bryggeri, 1959), 173.
13. E. C. Dahls bryggeri, *Aksjeselskapet E. C. Dahls bryggeri 1856–1956: 100 år* (Trondheim: E.C. Dahl, 1956), 104-105.
14. This example is taken from the Frydenlund brewery, quoted in Bryggeri- og mineralvannfabrikkarbeidernes forening, *"Vårt bidrag": Bryggeri- og mineralvannfabrikkarbeidernes forening gjennom 100 år: 1884–1984* (Oslo: Bryggeri- og mineralvannfabrikkarbeidernes forening, 1984), 27. The number of bottles they could wash per day was not given. Thirty øre in 1899 is the equivalent of 17.6 kroner in 2008 (approximately US$2.90). The source of the price conversion is Statistics Norway's online consumption price index calculator at http://ssb.no/vis/kpi/kpiregn.html. The average daily salary for working women in 1900 was about 1.3 kroner: Statistics Norway, "Historical statistics: Daglønn, etter yrke. Bygder og byer. 1875–1920. Kr," http://www.ssb.no/histstat/aarbok/ht-0605-257.html.
15. Mentz Schulerud, *Ringnes bryggeri gjennom 100 år* (Oslo: Rignes, 1976), 88.
16. Bottle cleaning also generated much waste. In a study from 1995, 30–50 percent of the wastewater generated in a brewery came from bottle washing activities. This water is highly caustic, with a pH level of 11.5. Norsk Bryggeri- og mineralvannsforening, *Miljøhåndbok for bryggeribransjen 1995* (Oslo: Norsk Bryggeri- og mineralvannsforening, 1995), 27.
17. E. C. Dahls bryggeri, *Aksjeselskapet E.C. Dahls bryggeri 1856–1956*, 104–105.
18. Øystå, *Brygg, brus og bruduljer*, 149; Chr. P. Killengreen, *Den Norske Bryggeriforening: Jubileumsskrift i anledning foreningens 25 aars bestaaen* (Oslo: Centraltrykkeriet, 1926), 25.
19. Øystå, *Brygg, brus og bruduljer*, 35.
20. The thirty-five- and seventy-centiliter sizes had been regulated by law since 1912.
20. Den Norske Bryggeriforening to Det kgl. Sosialdepartement, January 13, 1971, subfolder "Korrespondanse fra Industriforbundet til Sosialdepartementet, 15. januar 1971," folder 1265 "Lov om engangsemballasje 1970–1974," PA-0636 Norges Industriforbund, the National Archival Services of Norway, Oslo. Subfolder hereafter cited as subfolder "Korrespondanse fra Industriforbundet til Sosialdepartementet."
22. Edwin F. Lowry, Thomas W. Fenner, and Rosemary M. Lowry, *Disposing of Nonreturnables: A Guide to Minimum Deposit Legislation* (Stanford, Calif.: Stanford Environmental Law Society, 1975), 1.
23. U.S. Resource Conservation Committee, *Briefing Book on Beverage Container Deposit Policy*, Background paper no. 1 (Washington, DC, 1977), 1.
24. William K. Shireman, *The CalPIRG-ELS Study Group Report on Cans and Bottle Bills* (Stanford, Calif.: Stanford Environmental Law Society, 1981).
25. For more on Keep America Beautiful, see Finn Arne Jørgensen, "Keep America Beautiful," in *Encyclopedia of American Environmental History*, ed. Kathleen Brosnan (New York: Facts on File, 2010) and Rothman, *The Greening of a Nation?*

26. H. Lanier Hickman, Jr., *American Alchemy: The History of Solid Waste Management in the United States* (Santa Barbara, Calif.: Forester Press, 2003), 385.
27. Rogers, *Gone Tomorrow*, 149.
28. *Heritage of Splendor*, film, produced for Keep America Beuatiful by Richfield Oil Corporation (Hollywood, Calif.: Higgins [Alfred] Productions, 1963). The film is downloadable under a Creative Commons license as part of the Prelinger Collection of the Internet Archive: http://www.archive.org/details/Heritage1963.
29. In addition, Hawaii implemented a bottle bill in 2002 as the first American bottle bill passed after the 1980s. As of 2001, proponents for new bottle bills legislation were campaigning in Arkansas, Illinois, Maryland, Tennessee, and West Virginia, as well as for a national bottle bill. Campaigns for expanding current legislation are running in Connecticut, Massachusetts, and New York. The Container Recycling Institute's Bottle Bill Resource Guide at http://www.bottlebill.org/legislation/usa.htm.
30. Gandy, *Recycling and the Politics of Urban Waste*, 77.
31. "What This State Needs Is a Good 5 cent Returnable Bottle," *The West Sider*, 23, September 1976, quoted in Gandy, *Recycling and the Politics of Urban Waste*.
32. Blumberg and Gottlieb, *War on Waste*, 226.
33. The EPA Resource Conservation Committee sorted the participants in a 1977 public meeting on bottle bills into pro and con categories. Those who supported mandatory deposit legislation "as an inevitable course of action" were the League of Woman Voters, Crusade for a Cleaner Environment, the National League of Cities, the Sierra Club, the Rhode Island Solid Waste Management Institute, Environmental Action, Friends of the Earth, Earth Alive, Society of American Travel Writers, the Department of Defense, and the Adolph Coors Corporation (the only brewery in favor). In the con category, opposing bottle bills, were the Continental Group; the National Soft Drink Association; the U.S. Brewers Association; the Food Marketing Institute; the Can Manufacturers Institute; the Reynolds Metal Corporation; the Glass Packaging Institute; the American Glass and Steel Institute; the Stone, Glass, and Clay Committee of AFL-CIO; Keep America Beautiful, Inc.; the Beverage Industry Recycling Program; the National Automatic Merchandising Association; the Beer Distributors Recycling Fund; the Society of the Plastics Industry; the United Steelworkers of America; and the National Association of Retail Grocers of the United States. U.S. Resource Conservation Committee, *Briefing Book*, 4–5.
34. Pendergrast, *For God, Country, and Coca-Cola*, 273.
35. Mineralvannindustriens landslag to Det kongelige sosialdepartement, January 7, 1971, subfolder "Korrespondanse fra Industriforbundet til Sosialdepartementet."
36. Notes from meeting of the Packaging Forum, January 21, 1971, arranged by Moss Glass Works. ASv/RM, 29/1-71, subfolder "Korrespondanse fra Industriforbundet til Sosialdepartementet."
37. Øystå, *Brygg, brus og bruduljer*, 150.
38. Nordic Council Recommendation no. 5, 1969.
39. The Council of Europe was founded in 1948 as a political organization working to achieve greater unity among its member organizations. In 1970, the council had eighteen member states.
40. With a few notable exceptions like Erik Dammann, who wrote a book titled *Fremtiden i våre hender: Om hva vi alle kan gjøre for å styre utviklingen mot en bedre verden* (Oslo: Gyldendal, 1972) and founded an organization with the same name (The Future in Our Hands). His organization aims to decrease consumption in the Western world and is currently one of the largest Norwegian NGOs devoted to social change.

41. Norwegian scholars have studied Årdal Verk extensively. Two main works are Rolv Petter Amdam, Dag Gjestland, and Andreas Hompland, eds., *Årdal: verket og bygda 1947–1997* (Oslo: Samlaget, 1997) and Kristin Asdal, "Politikkens teknologier: Produksjoner av regjerlig natur" (PhD diss., University of Oslo, 2004).
42. Bredo Berntsen, *Grønne linjer: Natur- og miljøvernets historie i Norge* (Oslo: Grøndal Dreyer, 1994).
43. Norway, *Lov om naturvern*, Ot. prp. nr. 65, 1968–69 (June 19, 1970).
44. Norway, *Lov om adgang til å forby bruken av visse slag engangsemballasje ved markedsføring av forbruksvarer*, Ot. prp. nr. 77, 1969–70.
45. Norway, *Lov om adgang til å forby bruken av visse slag engangsemballasje*. The committee proposal really did claim that cans would eventually degrade, in contrast with glass bottles, which would not—a quite bizarre position considering that the environmental problem they were discussing was visual littering.
46. Norway, Odelsting discussion, "Diskusjon av Ot. prp. nr. 77, 1969–70," O. tid., 485–87, 1970.
47. "Helsedirektoratet ble utmanøvrert: Ølboks-saken lovkomedie," *Morgenposten*, July 31, 1970, and "Forsvinner øl på boks?" *Dagbladet*, August 5, 1970.
48. Norway, Odelsting discussion, "Diskusjon av Ot. prp. nr. 77, 1969–70."
49. Mineralvannfabrikantenes Landsforening to Industridepartementet, December 29, 1970, subfolder "Korrespondanse fra Industriforbundet til Sosialdepartementet."
50. Mineralvannfabrikkenes Landsforening to Det kgl Sosialdepartement, May 28, 1970, subfolder "Korrespondanse fra Industriforbundet til Sosialdepartementet."
51. Noblikk-Sannem AS, Norske Blikkemballasjefabrikker to Det kgl. Sosialdepartement, June 30, 1970, subfolder "Korrespondanse fra Industriforbundet til Sosialdepartementet."
52. Norges Handelsstands Forbund to Det kgl. Sosialdepartement, February 26, 1971, subfolder "Korrespondanse fra Industriforbundet til Sosialdepartementet."
53. Norsk Hydro A.S. to Industriforbundet, January 6, 1971, subfolder "Korrespondanse fra Industriforbundet til Sosialdepartementet."
54. SkanAluminium, Skandinavisk organ for aluminiumsindustrien, to Industriforbundet, February 22, 1971, subfolder "Korrespondanse fra Industriforbundet til Sosialdepartementet."
55. Elkem AS to Norges Industriforbund, August 19, 1970, subfolder "Korrespondanse fra Industriforbundet til Sosialdepartementet."
56. A/S Moss Glasværk to Norges Industriforbund, February 24, 1971, subfolder "Korrespondanse fra Industriforbundet til Sosialdepartementet."
57. Arne Ødegaard, Jan Grønli, and Ole-Jørgen Lier, *PLM Moss glassverk AS 1898–1998* (Moss: Glassverket, 1998).
58. Packaging Association to executive officer Melleby in the Ministry of Industry, February 9 (the paper is not marked with a year, but it is probably 1972, 1973, or 1974, as the containing folder has documents from 1970–1974, and the paper mentions documents from January 1972), folder 1265, "Lov om engangsemballasje 1970–1974," PA-0636 Norges Industriforbund, the National Archival Services of Norway, Oslo.
59. See Rothman, *The Greening of a Nation?* and Finn Arne Jørgensen, "Keep America Beautiful."
60. "Vedr. Emballasje," Notat, ASv/RM, 22/9-71, folder 1265, "Lov om engangsemballasje 1970–1974," PA-0636 Norges Industriforbund, the National Archival Services of Norway, Oslo.
61. Industriforbundet to Det kgl sosialdepartement, February 27, 1971, subfolder "Korrespondanse fra Industriforbundet til Sosialdepartementet."

62. Eivind Erichsen, "Må vi velge mellom økonomisk vekst og miljøvern?" in *Økonomi og politikk: 15 artikler*, ed. Petter Jakob Bjerve and Ole Myrvoll, 51–66 (Oslo: Aschehoug, 1971).
63. *Statsbudsjettet*, St. prp. no. 1, 1973–74, 6.
64. Kristin Asdal, *Økonomer og miljøavgifter—en historisk analyse*, Rapportserie fra Alternativ Framtid nr. 6 (Oslo: Alternativ Framtid, 1995), 15.
65. Norway, *Lov om produktkontroll*, Ot. prp. nr. 51, 1974–75 (April 18, 1975).
66. Norway, *Statsbudsjettet*, St. prp. no. 1, 1973–74, 6.
67. Norway, *Om midlertidig lov om panteordninger for emballasje til øl, mineralvann og andre leskedrikker*, Ot. prp. nr. 61, 1973–74 (May 27, 1974).
68. Norway, Storting discussion, "Innstilling fra finanskomiteen om midlertidig lov om panteordninger for emballasje til øl, mineralvann og andre leskedrikker" (innst. O. Nr. 57), S. tid. 651, 1974.
69. Norway, Ministry of Finance, "Forskrifter om panteordninger for emballasje til øl, mineralvann og andre leskedrikker. Forskrifter om avgift på øl og kullsyreholdige, alkoholfrie drikkevarer i engangsemballasje," June 14, 1974.
70. Norway, Ministry of the Environment, *Resirkulering og avfallsbehandling II*, NOU 1975:52, 119.

Chapter 3

1. The Planke brothers have told and retold the story of how the first RVM was installed many times in slightly different versions. The version I have told is based on Carol Quinn's interviews with the Planke brothers, as well as my own interviews with the Plankes and Aage Fremstad. Carol Quinn interviewed Tore and Petter Planke on behalf of Tomra in 2000; videotapes of the interviews are held in the Tomra corporate archives.
2. SINTEF later changed to its current name, the Foundation for Scientific and Industrial Research.
3. The description of bottle handling in grocery stores is based on interviews with Aage Fremstad and the Planke brothers, newspaper articles, and Erik Røsrud's history of Norwegian grocery stores, *Dagligvareforbundets rolle og virke mot år 2000* (Oslo: Dagligvareforbundet, 2003).
4. For more on vending machines, see G. R. Schreiber, *A Concise History of Vending in the U.S.A.* (Chicago: Vend, 1961); Karl H. Ulleberg, *Automathandelen i Norge og automater som omsetningsform* (Bekkestua: Norges Handelshøyskole, 1968); and Kerry Segrave, *Vending Machines: An American Social History* (Jefferson, N. C.: McFarland, 2002).
5. "Salgs-automatene får stadig økende utbredelse" *Kjøpmannsnytt* 45, no. 3 (1967): 41.
6. The first patent application to use the term *reverse vending machine* was filed by Jim M. Swendeman (assigned to Reynolds Metal Company) in 1982, US patent no. 4,432,279. I have not been able to determine whether the term was in use before then. In older patent applications, the machines are called bottle/can/beverage container collection/return/handling machines. I alternate between using the terms *reverse vending machine*, *RVM*, and *bottle return machine* when referring to the machine. In the first English-language Tomra annual report from 1984, they used the term *reverse vending equipment*.
7. Aktielaget Wicanders Korkfabriker to Arthur Tveitan AS concerning Aage Fremstad, October 2, 1962. Aage Fremstad private collection.
8. Aage Fremstad, interviewed by author, Oslo, Norway, August 1, 2005, digital recording.
9. "Tomras ukjente konkurrent," *Økonomisk Rapport*, no. 19 (1990): 12.

10. Fredrik Tveitan, interviewed by author, August 15, 2005, phone interview, written notes.
11. Gunnar C. Aakvaag, *Forbrukersamvirket og medlemmene 1970–2004. Mellom sosialdemokratisk modernisering og nyliberal individualisering*, ISF Report 2004:18 (Oslo: Institutt for samfunnsforskning, 2004), 15.
12. Fredrik Tveitan, interviewed by author.
13. "Tomras ukjente konkurrent," *Økonomisk Rapport*, no. 19 (1990): 12.
14. "Teknisk komite—Mottak av tomflasker i forretningene," April 20, 1971. Aage Fremstad private collection.
15. One of my colleagues reminisced about how as a kid he used to take *full* bottles from the store shelves and insert them in the bottle return machine, which at the time could not measure the weight of the bottles, only the size.
16. "En viktig faktor i returflaskeautomater: Det er mekanikken som selger," *Elektro*, October 4, 1979, 32.
17. Aage Fremstad, interviewed by author.
18. Arvid Strand, *Dagligvarekjøpmennene og samfunnet 1958–1988* (Oslo: Norges dagligvarehandels forbund, 1988), 401.
19. "Oppsummering og vurdering av de argumenter som er framkommet fra bryggeriene for og mot forhöyelse av panten for 1/1-flasker for öl til 60 öre," April 19, 1971. Aage Fremstad private collection.
20. Oslo Kolonialkjøpmenns Forening, "Ingen løsning på flaskepantsaken," Sirk. Nr. 7/71, July 21, 1971. Aage Fremstad private collection.
21. "Teknisk komite—Mottak av tomflasker i forretningene," February 17, 1971. Aage Fremstad private collection.
22. "Teknisk komite—Mottak av tomflasker i forretningene," February 17, 1971. Aage Fremstad private collection.
23. "Teknisk komite—Mottak av tomflasker i forretningene," April 14, 1971. Aage Fremstad private collection.
24. Wiebe Bijker argues that many inventions often come from outside the established way of thinking about the problem in *Bicycles, Bakelites, and Bulbs: Toward a Theory of Sociotechnical Change* (Cambridge: MIT Press, 1995). His concept of technological frames illustrates how interest groups operate within established ways of conceptualizing problems and solutions. Outside actors often bring in new knowledge and new ways of seeing the problem.
25. Petter Planke, interviewed by author, Vollen, Norway, August 2, 2005, digital recording. This is another fragment of Tomra's history that has been told and retold so many times that the exact story may have changed over the years.
26. Sverre Martens Planke, *Det var en gang: Erindringer fra et helt århundre* (Tvedestrand: Toring AS, 2001), 241.
27. The biographical information on Petter Planke is based interviews with him, Sverre M. Planke's autobiography, as well as Stian Jacobsen, *Norges beste gründere—og floppene* (Oslo: Hegnar Media, 2002), 209.
28. The biographical information on Tore Planke is based on interviews with him and Sverre M. Planke's autobiography. The information on his thesis comes from an email from Tore Planke to Finn Arne Jørgensen, November 13, 2006.
29. Tore Planke, "Apparat for automatisk mønstergjenkjenning og registering av tomflasker," Norwegian patent no. 126900, filed December 14, 1971, and patented April 9, 1973.
30. Norwegian patent no. 126900, 2.
31. "Flinke folk er billige i Norge," *Økonomisk Rapport*, no. 17 (1983): 29.
32. For a good discussion of inventors' use of patents, see Chapter 2 in Hughes, *American Genesis*. Although Tore did find a few patents at the time, years later he

discovered that there were in fact more than twenty RVM patents prior to 1971 worldwide.

33. "En viktig faktor i returflaskeautomater: Det er mekanikken som selger," in *Elektro*, October 4, 1979, 32.
34. Elmer M. Jones, "Vending Machine," US patent no. 1,560,242, filed September 13, 1920, and issued November 3, 1925.
35. Samuel J. Gurewitz, "Bottle Handling Machine," US patent no. 2,750,024, filed February 19, 1952, and issued June 12, 1956, and Samuel J. Gurewitz, "Bottle Register," US patent no. 2,908,440, filed June 19, 1953, and issued October 13, 1959.
36. Bruce Garrard, "Bottle Vending Machine," US patent no. 2,804,958, filed January 25, 1954, and issued September 3, 1957.
37. Samuel J. Gurewitz, "Bottle Register," 7.
38. Tore Planke, interviewed by Carol Quinn, August 8, 2000, Asker, Norway, video recording, tape 1, Tomra corporate archives.
39. "Teknisk komite—Mottak av tomflasker," December 21, 1971. Aage Fremstad private collection.
40. This amounts to 115,000 kroner ($18,500) converted to 2007 values.
41. "Sluttrapport fra teknisk komite," January 19, 1972. Aage Fremstad private collection.
42. However, some stores received even more. On Saturdays, large stores like the EPA shopping center in Oslo could receive up to ten thousand bottles in just one day. At these times, the customers had to wait in line for half an hour to return their bottles. Petter Planke, interviewed by author, Vollen, Norway, August 2, 2005, digital recording.
43. "Sluttrapport fra teknisk komite," January 19, 1972.
44. Oslo Kolonialkjøpmenns Forening, "Tomflaskeautomat," January 25, 1972. Aage Fremstad private collection.
45. Tore and Petter Planke, interviewed by Carol Quinn, Asker, Norway, August 8, 2000, video recording, tape 1, Tomra corporate archives.
46. Some examples: "Ingen pant—ingen retur," *Dagens Nyheter*, July 4, 1970; and "Hold naturen ren," *Norges Kjøbmannsblad*, June 13, 1970.
47. The Planke brothers had participated in a marketing course where the Kodak story was mentioned. For more on the naming of Kodak, see Susannah Hart and John Murphy, eds., *Brands: The New Wealth Creators* (Basingstoke, U.K.: Palgrave, 1998), 42.
48. They would later discover that their name was not quite as successful as Kodak, since Tomra often became "Tumra" in the United States and "Tomla" in Japan.
49. Norcontrol has been studied by Signy Ryther Overbye, "Etableringen av en norsk skipsautomatiseringsindustri," in *Elektronikkentreprenørene: Studier av norsk elektronikkforskning og –industri etter 1945*, ed. Olav Wicken, 152–177 (Oslo: Ad Notam Gyldendal, 1994).
50. Sverre M. Planke, *Det var en gang . . .*, 245.
51. Tomra annual report, 1972.
52. Tomra annual report, 1972. This was just a fraction of the total number of bottles returned—in 1973, 825 million bottles were returned. Norway, Ministry of the Environment, *Resirkulering og avfallsbehandling II*, NOU 1975:52, 119.
53. In actuality, the number of bottles per day was higher because just a few machines were installed at the beginning of the year.
54. "Automatic Bottle Collector," Tomra advertising brochure, 1974, 7. Tomra corporate archives.
55. Tore Planke, interviewed by author, Oslo, Norway, January 17, 2006, digital recording.

56. Tomra annual reports, 1973 and 1974.
57. Tomra annual report, 1973, 5.
58. Petter Planke, interviewed by author, Vollen, Norway, August 2, 2005, digital recording.

Chapter 4

1. Tore Planke, interviewed by Carol Quinn, August 8, 2000, Asker, Norway, video recording, tape 3, Tomra corporate archives.
2. Adjusted to 2009 prices, the Tomra I cost about 145,000 kroner ($23,500).
3. I have not found any price data for the Tveitan RVM, and Fredrik Tveitan did not remember when I interviewed him. However, when Tomra made their Junior model to compete with Tveitan, they priced it at approximately twelve thousand kroner.
4. Tveitan ATAS advertising brochure, not dated. Fredrik Tveitan private collection.
5. Tomra annual report, 1974, 3.
6. Tore Planke, interviewed by Carol Quinn, August 9, 2000, Asker, Norway, video recording, tape 7, Tomra corporate archives.
7. Sweden, Riksdagen, "Avgift på vissa dryckesförpackningar," Prot. 1973:27, February 21, 1973.
8. Sweden, Ministry of Heath and Social Affairs, *Ett renare samhälle*, SOU 1969: 18, 15.
9. OECD, *Beverage Containers—Re-use or Recycle* (Paris: Organization for Economic Co-operation and Development, 1978), 22. The öre is the smallest Swedish monetary unit. There are one hundred öre to one krona.
10. Sweden, Ministry of Heath and Social Affairs, *Ett renare samhälle*, 31.
11. The Rigello bottle was developed by the packaging company Tetra Pak in cooperation with the Pripps brewery in the second half of the 1960s. While technically a disposable container, Rigello Pak strongly highlighted the environmental friendliness of the lightweight Rigello bottle, particularly through a series of publications with technical information on waste disposal, energy consumption, and recyclability. See Rigello Pak AB, *Basfakta i debatten miljövård—dryckesförpackningar* (Lund: Rigello Pak, 1969); Rigello Pak AB, *Plastics from an Environmental Standpoint: Kunststoffe und unsere Umwelt = Matières plastiques et l'environnement* (Lund: Rigello Pak, 1970); Rigello Pak AB, *Dryckesförpackningarna och energin = Beverage containers and energy = Getränkeverpackungen und die Energie = Emballages de boisson et consommation d'énergie* (Lund: Rigello Pak, 1974).
12. Karl Lidgren, *Dryckesförpackningar och miljöpolitik—en studie av styrmedel* (Lund: Lund Economic Studies, 1978), 10. The number of new beverage containers produced every year in Sweden was thus almost as large as the *total* number of beverage containers in use in Norway.
13. Tomra annual report, 1973, 3.
14. Petter Planke gleefully retold the story of how during the first test run in Sweden they observed how the customers used the machine. One elderly lady came in and began pulling gin bottles out of her bag—first one, then two, and then a third one, before she noticed that she was being observed. Turning bright red, she continued inserting the bottles until she came to bottle number eight. At that point she turned to the observers and said: "It's for my son!" Experiences like this illustrate how the bottle machines intersected in consumers' private lives. When bottles were returned manually, the store personnel could see exactly what you had drunk. With the machines, this transaction could be more anonymous. The old lady's reaction shows us what happens when this anonymity disappears. Petter Planke,

interviewed by Carol Quinn, Asker, Norway, August 8, 2000, video recording, tape 1, Tomra corporate archives.

15. "Bedriften som lever av tomflasker," *Elektronikk*, no. 7 (1977): 9.
16. Tomra annual report, 1972, 2.
17. OECD, *Beverage Containers*, 26.
18. Andreas Golding, *Reuse of Primary Packaging, Final Report*, study B4–3040/98/0001 080/MAR/E3 (Brussels: European Commission Environment Directorate-General, 1998), France country report, http://ec.europa.eu/environment/waste/studies/packaging/pdf/france.pdf
19. Tore and Petter Planke, interviewed by Carol Quinn, Asker, Norway, August 8, 2000, video recording, tape 3, Tomra corporate archives.
20. The first-generation Tomra machines were labeled I, II, or III, combined with the letters A, B, or C. The numbers indicated the frame type and how it was welded together, whereas the letters reflected the arrangement of the optics.
21. "A/S Tomra Systems med nytt 'skudd': Ny elektronikkbedrift etablert i Trondheim," *Norges Industri*, May 30, 1983, 7.
22. Tore Planke, interviewed by Carol Quinn, August 8, 2000.
23. Tore and Petter Planke, interviewed by Carol Quinn, August 9, 2000.
24. Tore Planke, interviewed by Carol Quinn, August 8, 2000.
25. "Billedbehandling og mønstergjenkjenning i Norge: Nye fagfelt gror i Norge," *Teknisk Ukeblad*, June 21, 1984, 12.
26. "Askerladdprisen gikk til Tomra," *Asker og Bærums Budstikke*, December 5, 1984.
27. Tomra annual report, 1975, 4.
28. Tomra annual report, 1976, 4.
29. See, for instance, "Returflaske-automat med solid forsprang," *Arbeiderbladet*, May 23, 1978.
30. Tomra annual report, 1977, 2–3.
31. Tomra annual report, 1980, 2.
32. "Elektronikk i mekaniske produkter: Et utfordrende felt med store muligheter," *Elektronikk*, no. 4A (1979): 19–20.
33. Tomra annual report, 1973.
34. The relationship between Tomra and Hugin, however, took a turn for the worse after Electrolux purchased Hugin. Upon evaluating Hugin's shares in Tomra, Electrolux decided that it liked the company so much that it wanted to also acquire a majority share in Tomra. The Planke brothers were not ready to sell, and they negotiated an expensive buyback of the Hugin shares. However, just six months later, Tomra sold the same shares—33 percent of the company—to external investors for three times the amount. From this point on, Tomra's financial worries were over for a few years.
35. Tore Planke later retold the story of how some DARPA (Defense Advanced Research Projects Agency) engineers visited Tomra in the early 1990s to take a look at the CRA recognition technology. They were stunned at the speed and precision of the CRA and of the fact that this was mass produced in a civilian technology. As they saw it, American military technology could not match the CRA. Tore Planke, interviewed by author, Asker, Norway, June 14, 2005. Digital recording.
36. The Tomra 300 did not launch until 1987. Bernt Rønningsbakk studied the development of the Tomra 300 RVM as one of the cases in his engineering dissertation. His description of the innovation process has been a valuable source in my discussion of the Tomra 300. See Bernt Rønningsbakk, "Nyskaping og dialog: Med 12 norske eksempler" (doctoral thesis, Norges Tekniske Høyskole, 1995).

37. Joseph A. Schumpeter, "Economic Theory and Entrepreneurial History," in *Essays of J. A. Schumpeter*, ed. R. V. Clemence, 248–66 (Cambridge, Mass.: Addison-Wesley Press, 1951).
38. Tore Planke, "Teknologien skyver på med stadig økende styrke," in Tomra annual report, 1983, 2.
39. Tomra annual report, 1985, 5.
40. Planke, "Teknologien skyver på med stadig økende styrke."
41. Norway, Ministry of the Environment, *Resirkulering og avfallsbehandling II*, NOU 1975:52, 117.
42. Jury chairman Anton Merckoll quoted in "Eksportproduktprisen til A/S Tomra Systems," *Asker og Bærums Budstikke*, April 1, 1984.
43. See, for instance, Truls Fallet, "Prisverdig," *Elektro*, April 14, 1981, 9.
44. "Millionærer på flaskepanting," *Nå*, March 2, 1983, 32.
45. Petter Planke, "Moderne elektronikk: Hvorfor—For hvem—Fra hvem—Hvordan?" lecture for NTNF/Datatid, October 4, 1980. Manuscript from Petter Planke's private collection.
46. "Elektronikkbransjen viktig for den videre industriutvikling," *Arbeidsgiveren*, February 4, 1982; "Vi må så før vi kan høste," *Elektro*, March 3, 1983, 9; Petter Planke, "Datateknologi og arbeidsplasser," *Aftenposten*, July 29, 1981; and "A/S Tomra Systems med nytt 'skudd': Ny elektronikkbedrift etablert i Trondheim," *Norges Industri*, May 30, 1983, 10.
47. "Lure ideer skaper ikke ny industri," *Arbeider-Avisa*, October 25, 1983.
48. Petter Planke, "Fri idéskapning som grunnlag for industriutvikling," *Teknisk Ukeblad*, no. 48A (1983): 10.
49. "A/S Tomra Systems med nytt 'skudd': Ny elektronikkbedrift etablert i Trondheim," *Norges Industri*, May 30, 1983, 9.
50. Planke, "Teknologien skyver på med stadig økende styrke."
51. Tomra annual report, 1983, 2.
52. Tomra annual report, 1983, 4.

Chapter 5

1. *Seinfeld*, "The Bottle Deposit," part 1 and 2, season 7, episode 21 and 22, NBC, aired May 2, 1996.
2. U.S. Department of Justice, "Ten Plead Guilty in Scheme That Defrauded California's Bottle/Can Recycling Program with Foreign and Non-existent Containers," Press Release 06-035, March 22, 2006.
3. Aluminum Company of America, *Aluminum by Alcoa* (Pittsburg, Pa.: Aluminum Company of America, 1969), 2.
4. Robert Friedel, "A New Metal! Aluminum in Its 19th-Century Context," in *Aluminum By Design*, ed. Sarah Nichols, 59–83 (Pittsburgh: Carnegie Museum of Art, 2000), 59.
5. United States Geological Service, *Aluminum Statistics: 1900 to 2002* (Denver, Colo.: USGS, August 2004), 2.
6. Thomas Hine, *The Total Package: The Secret History and Hidden Meanings of Boxes, Bottles, Cans, and Other Persuasive Containers* (Boston: Little, Brown, 1997), 161.
7. Baum, *Citizen Coors*, 38.
8. George David Smith, *From Monopoly to Competition: The Transformations of Alcoa, 1888–1986* (Cambridge: Cambridge University Press, 1986), 389.
9. John T. Aquino, *Waste Age / Recycling Times' Recycling Handbook* (Washington, DC: Environmental Industry Associations, 1995), 13.

10. Aquino, *Waste Age*, 13.
11. Sweden, Ministry of the Environment, *Översyn av skatten på dryckesförpackningar*, SOU 1990:85, 37.
12. "Plåtmanufaktur" translates directly to "Sheet Manufacturing." PLM is today part of Rexam, one of the world's largest consumer packaging manufacturers.
13. PLM was not first to consider aluminum cans in Sweden, though. The smaller Wårby Bryggeriar made plans for aluminum cans in the mid-1960s, but the Swedish Environmental Protection Agency convinced Wårby to not go ahead with the plans. Sweden, Riksdagen, "Om tillverkningen av öl- och läskburkar av aluminium," Prot. 1979/80:42, December 3, 1979, 81–83.
14. Harald Biong, *Mulige konsekvenser av å innføre aluminiumsboks som drikkevareemballasje i Norge i forbindelse med et pante- og retursystem. Prosjekt Del I – En studie av aluminiumsboksens utvikling i Sverige* (Sandvika: Norsk Institutt for Markedsforskning, 1990), 14.
15. Sweden, Ministry of the Environment, *Översyn av skatten på dryckesförpackningar*, 25.
16. Sweden, Riksdagen, "Om PLMs tillverkning av aluminiumburkar i Fosie," Prot. 1980/81:94, March 12, 1981.
17. Sweden, *Lag om återvinning av dryckesförpackningar av aluminium*, 1982:349.
18. Sweden, Riksdagen, *Regeringens proposition om återvinning och omhändertagande av avfall*, Proposition 1975:32 (February 27, 1975).
19. Sweden, Tax Committee, "Om återvinning av dryckesförpackningar av aluminium (prop. 1981/82:131)," SkU 1981/82:61, 3.
20. Sweden, Ministry of Agriculture, *Utredning om frågor om återvinning av dryckesförpackninger*, DS Jo 1980:05.
21. Tore and Petter Planke, interviewed by Carol Quinn, August 8, 2000, video recording, tape 3, Tomra corporate archives.
22. Sweden, Ministry of Agriculture, *Dryckesförpackningar och miljö*, SOU 1976:35, 10.
23. Sweden, Ministry of Heath and Social Affairs, *Ett renare samhälle*, 141.
24. "Test med al-burk i retur på Gotland: Automaterna räckte inte," *FK*, November 26, 1982.
25. "Daro—konkurrent til stora burkpressföretag," *Ottens-kuriren*, May 20, 1983.
26. "1 mars börjar burk-panten att gälla," *Handelsnytt*, nd. (January or February 1983). Copy found in Tomra's newspaper clippings archive.
27. Sweden, Riksdagen, "Svar på fråga 1984/85:384 om insamlingen av aluminiumburkar," Prot. 1984/85:87, February 21, 1985, 100.
28. Tore and Petter Planke, interviewed by Carol Quinn, August 8, 2000, video recording, tape 3, Tomra corporate archives.
29. Tomra added a barcode reader when they introduced Can-Can to the American market.
30. "Hjemmesløyd selger ikke," *Økonomisk Rapport*, no. 19 (1983): 33.
31. Petter Planke, interviewed by author, August 2, 2005, digital recording.
32. "Askerbedrift hedres torsdag: Design-pris til Tomras Can-can," *Asker og Bærums Budstikke*, April 4, 1984.
33. "Tomra hedret for 'God norsk design'," *Asker og Bærums Budstikke*, April 10, 1984.
34. The council's predecessor, the Norwegian Design Centre, awarded the Mark for Good Design from 1965 to 1973. The 1984 award was the first time the Council for Industrial Design gave out this award. See Kjetil Fallan, "How an Excavator Got Aesthetic Pretensions—Negotiating Design in 1960s' Norway," *Journal of Design History* 20, no. 1 (2007), 46.
35. Tore and Petter Planke, interviewed by Carol Quinn, August 8, 2000, video recording, tape 3, Tomra corporate archives.
36. Sweden, Riksdagen, *Om återvinning av dryckesförpackningar av aluminium*, Proposition 1981/82:131 (February 24, 1982), 9.

37. Sweden, Riksdagen, *Om återvinning av dryckesförpackningar av aluminium*, Proposition 1981/82:131 (February 24, 1982), 9.
38. Tomra annual report, 1983, 5.
39. PLM won orders for six hundred to seven hundred RVMs. Tore and Petter Planke, interviewed by Carol Quinn, August 8, 2000, video recording, tape 3, Tomra corporate archives.
40. Tore and Petter Planke, interviewed by Carol Quinn, August 8, 2000, video recording, tape 3, Tomra corporate archives; Tomra annual report, 1984, 5. According to the 1983 annual report, only sixteen hundred machines were actually in place by March 1, when they were all turned on. This illustrates the challenges of relying on memory and oral testimony when constructing historical narratives.
41. Tore Planke, interviewed by Carol Quinn, August 8, 2000, video recording, tape 3, Tomra corporate archives. Evidence of this can be found in Tomra's newspaper clippings folders. Two thick folders are full of articles about Tomra's machines—a large proportion of them are negative.
42. This company later changed its name to Scandinavian Industries.
43. Gurewitz, "Bottle Handling Machine," US patent no. 2,750,024.
44. Later, SMI changed its name to FMI, Food Marketing Institute.
45. Tore and Petter Planke, interviewed by Carol Quinn, August 8, 2000, video recording, tape 3, Tomra corporate archives.
46. As previously mentioned, these were Michigan and Maine in 1976, Iowa and Connecticut in 1978, Delaware and New York in 1982, and Massachusetts in 1983.
47. L. B. McEwen, *Waste Reduction and Resource Recovery Activities: A Nationwide Survey* (Washington, DC: US Environmental Protection Agency, 1977)
48. *Newsweek*, June 14, 1971, quoted in Oregon Department of Environmental Equality, "Oregon Bottle Bill—Then and Now," http://www.deq.state.or.us/lq/sw/bottlebill/thenandnow.htm.
49. Brent Walth, *Fire at Eden's Gate: Tom McCall and the Oregon Story* (Portland: Oregon Historical Society Press, 1994), 320–322.
50. Ackerman, *Why Do We Recycle?* 126.
51. Kenneth A. Gould, Allan Schnaiberg, and Adam S. Weinberg, *Local Environmental Struggles: Citizen Activism in the Treadmill of Production* (Cambridge: Cambridge University Press, 1996), 127.
52. Supposedly the barge was full of hazardous materials, but when the contents were finally burned in Robert Moses's South Brooklyn incinerator, nothing out of the ordinary was found—just regular household and industrial waste. The story has been described in, e.g., Blumberg and Gottlieb, *War on Waste* and Miller, *Fat of the Land*.
53. Subtitle D (added 1991) of the Resource Conservation and Recovery Act of 1976 specified minimum standards for all solid waste landfills in the United States, including composite liner systems to prevent seepage: Aarne Vesilind and Thomas Distefano, *Controlling Environmental Pollution: An Introduction to the Technologies, History, and Ethics* (Lancaster, PA: DEStech Publications, 2005), 311.
54. Ackerman, *Why Do We Recycle?* 1.
55. "Tomraavtale i USA," *Aftenposten*, May 10, 1984, morning edition.
56. "Norsk firma markedsleder i ny teknologisektor," *Teknisk Ukeblad*, July 26, 1984, 11.
57. Of the about twenty can machine producers in the United States at the time, Envipco was the largest. This company competed with Tomra in Scandinavia for a period in the early 1980s, distributed Tomra machines in the United States after the Tomra's 1986 crisis, and is still one of Tomra's largest global competitors.
58. "Tomra i teten: Automat som også tar plastbokser," *Aftenposten*, March 6, 1985, evening edition.

59. *Tomra Can-Can*, promotional video (Asker, Norway: Tomra Systems, 1984).
60. Tomra annual report, 1984, 4.
61. "Tomra Systems etablerer seg i USA, Canada og Nederland," *Eksport aktuelt*, January 25, 1984, 1.
62. "Tomra stiger mot nye høyder," *Aftenposten*, March 8, 1984, morning edition.
63. "Tomra inngår samarbeidsavtale i USA—har utviklet laser-skanner for lesing av strek-koder," *Fritt Kjøpmansskap*, May 1984.
64. "Tomraavtale i USA," *Aftenposten*, May 10, 1984, morning edition.
65. "Mulig storordre, men bare en begynnelse," *Asker og Bærums budstikke*, May 2, 1984; "Kjempeordre til Tomra," *Norges Handels og Sjøfartstidende*, May 2, 1984; and "Helspent Planke," *Norges Handels og Sjøfartstidende*, October 22, 1985.
66. "A Hardship on Deposit," *New York Times*, November 8, 1981.
67. See for instance Ackerman, *Why Do We Recycle?* 126.
68. "Issue and Debate; Who Gets 5¢ Deposits Unclaimed by Buyers," *New York Times*, March 18, 1985.
69. G. Oliver Koppell, *Inconspicuous Consumption: A Look at New York's Bottle Bill and the Experience of Six States, Which Have Enacted Returnable Container Legislation* (Albany, NY: New York State Assembly, 1981).
70. "Beer Prices Soar under NY Deposit Law," *New York Times*, September 17, 1983.
71. "Brands Deflected by Bottle Law Return," *New York Times*, November 2, 1983.
72. "State Bottle Law Upsets Industry," *New York Times*, July 12, 1982.
73. In 1989, the *New York Times* reported that many of the city's poorest could earn up to thirty-five dollars for eight hours work collecting cans and bottles: "Can Picker: $35 a Shift, No Benefits, No Bosses," September 6, 1989. For more descriptions of the importance of the beverage container deposit for New York's poor and homeless, see also "Resurgence of Aluminum Recycling Brings Cash to the Poor," August 8, 1983; "Cashing in Bottles, Living on Nickels," December 28, 1986; and "A Middleman's Adventures in the Can Trade," September 23, 1992, A1.
74. "Cashing in Bottles, Living on Nickels," *New York Times*, December 28, 1986.
75. "Cans, a Man, a Plan: New Redemption Center Struggles," *New York Times*, November 4, 1987.
76. Tore and Petter Planke, interviewed by Carol Quinn, August 8, 2000, video recording, tape 3, Tomra corporate archives.
77. City of New York Department of Sanitation, *Public Comments on a Comprehensive Solid Waste Management Plan for New York City and Draft Generic Environmental Impact Statement Submitted in June 1992* (New York: City of New York Department of Sanitation, 1992).
78. "Machines Will Trade Nickels for Beverage Cans," *New York Times*, September 12, 1983.
79. "Godt, men ikke godt nok for Tomra-gruppen i 1985," *Gjengangeren*, March 7, 1986.
80. Ackerman, *Why Do We Recycle?* 132.
81. "Issue and Debate; Who Gets 5¢ Deposits Unclaimed by Buyers," *New York Times*, March 18, 1985.
82. "Tomra i teten: Automat som også tar plastbokser," *Aftenposten*, March 6, 1985, evening edition.
83. Tore and Petter Planke, interviewed by Carol Quinn, August 8, 2000. Video recording, tape 4.
84. Metropolitan Mining remained in business, though, and was purchased by Envipco in 2001.
85. Gandy, *Recycling and the Politics of Urban Waste*, 78. The information comes from a report on the Moreland Act Commission on the Returnable Container Act submitted in March 1990 to Governor Mario Cuomo of New York State.

86. Blumberg and Gottlieb, *War on Waste*, 226.
87. "Kursene har steget under Tomra-tegning," *Aftenposten*, December 16, 1983, morning edition.
88. "De nye selskapene," *Aftenposten*, March 31, 1984, morning edition.
89. "Ukens aksje: Tomra Systems," *Tromsø*, November 25, 1985.
90. "Dårlig halvårsresultat fra Tomra: Var det ikke plankekjøring?" *Kapital*, no. 18 (1985).
91. "Helspent Planke," *Norges Handels og Sjøfartstidende*, October 22, 1985.
92. "Inntjeningssvikt for Tomra-gruppen," *Asker og Bærums Budstikke*, February 10, 1986.
93. "Inntjeningssvikt for Tomra-gruppen," *Asker og Bærums Budstikke*, February 10, 1986.
94. "Dårlig halvårsresultat fra Tomra: Var det ikke plankekjøring?" *Kapital*, no. 18 (1985).

Chapter 6

1. Tomra annual report, 1989, 4.
2. Tomra annual report, 1989, 22.
3. World Commission on Environment and Development, *Our Common Future* (New York: Oxford University Press, 1987); Braden R. Allenby and Deanna J. Richards, eds., *The Greening of Industrial Ecosystems* (Washington, DC: National Academy Press, 1994); Andrew Jamison, *The Making of Green Knowledge: Environmental Politics and Cultural Transformation* (Cambridge: Cambridge University Press, 2001); Árni Sverrison, "Translation Networks, Knowledge Brokers, and Novelty Construction: Pragmatic Environmentalism in Sweden," *Acta Sociologica* (2001): 313–327; and Maarten A. Hajer, *The Politics of Environmental Discourse: Ecological Modernization and the Policy Press* (Oxford: Clarendon Press, 1995).
4. Joseph A. Schumpeter, *The Theory of Economic Development: An Inquiry into Profits, Capiral, Credit, Interest, and the Business Cycle* (Cambridge: Harvard University Press, 1934), 93.
5. Schumpeter, *The Theory of Economic Development*, 78.
6. See Hughes, *Networks of Power*.
7. Michael J. Lynskey, introduction to *Entrepreneurship and Organization: The Role of the Entrepreneur in Organizational Innovation*, ed. Michael J. Lynskey and Seiichiro Yonekura, 1–57 (Oxford: Oxford University Press, 2002), 28; also referred to by Alfred D. Chandler, Jr., *Strategy and Structure: Chapters in the History of the Industrial Enterprise* (Cambridge: MIT Press, 1962), 12.
8. "Kritikk på Tomras generalforsamling: Styret sov for lenge," *Aftenposten*, September 19, 1986; see also "Tomra refinansieres: Massiv kritikk av styre og ledelse," *Aftenposten*, September 11, 1986.
9. "Kryssfinér," *Norges Handels og Sjøfartstidende*, August 1, 1986.
10. "-Næringslivet oser av feighet," *Ledelse*, no. 15 (1988): 30.
11. "Tomra-sjefen: Næringslivets Rambo?" *Ledelse*, no. 10 (1987): 4.
12. "Kryssfinér," *Norges Handels og Sjøfartstidende*, August 1, 1986.
13. "Tomra i USA: Derfor gikk det galt," *Økonomisk Rapport*, no. 15 (1986): 24.
14. "Tomra i USA," 26.
15. "Tomra forblir uavhengig," *Økonomisk Rapport*, no. 15 (1986): 25.
16. Tomra annual report, 1974, 1
17. Tore Planke, interviewed by author, Oslo, Norway, January 17, 2006, digital recording.

18. Tore Planke interviewed on the CNN program *Agenda Earth* in 1992 (no date), courtesy of Tomra archives.
19. "Vil ri på miljøbølgen," *Kapital Data*, no. 8 (1988).
20. "Ordholden revisor," *Dagens Næringsliv*, March 8, 1991.
21. For a discussion of the green wave, see Steven Yearley, *The Green Case: A Sociology of Environmental Issues, Arguments, and Politics* (London: HarperCollinsAcademic, 1991).
22. "- Lovende, ny miljøbevissthet," *Aftenposten*, February 13, 1989.
23. "Positiv miljøstrid," *Aftenposten*, March 31, 1989.
24. Berntsen, *Grønne linjer*.
25. Mission statement of 1914 quoted in Berntsen, *Grønne Linjer*.
26. "Reklamen blir grønn," *Dagens Næringsliv*, July 28, 1989.
27. "Én million tenker grønt," *Aftenposten*, October 30, 1989. How real were the anecdotal impressions about the green wave? Both Norsk Monitor and Norwegian Social Science Data Services performed a series of consumer surveys that demonstrate the extent of the green wave in the 1980s and 1990s. The surveys show that consumers increasingly preferred environmental protection to economic growth and industrial interests throughout the 1980s. In the 1990s, however, materialism, individualism, comfort, and convenience gradually caused the consumer to take on an increasingly lighter shade of green. Despite widespread environmental consciousness, actual consumer behavior did not always match the good intentions. Most people said that they had continued buying particular products after being told that the producer was a polluter. Comfort and convenience turned out to be more important than the environment for consumers in the end. For a more in-depth discussion of these surveys, see Knut Kalgraff Skjåk and Bjug Bøyum, *Undersøking om verdier, natur og miljø 1993* (Oslo: Norsk Samfunnsvitenskapelig Datatjeneste, 1993) and Kristin Stølsbotn, *Undersøkelse om verdier, natur og miljø 2000* (Oslo: Norsk Samfunnsvitenskapelig Datatjeneste, 2002).
28. "- Avfall kan behandles mye bedre," *Aftenposten*, July 6, 1989.
29. " Fra Ng til G+ i miljøvern," *Aftenposten*, July 12, 1990.
30. "Kjøpmann på den grønne vei," *Aftenposten*, September 15, 1988.
31. "På tomannshånd," *Aftenposten*, January 14, 1989.
32. "Miljøbevegelsen avslører: - Grønne varer er bevisst lureri," *Aftenposten*, May 8, 1991.
33. "Grønn bølge i markedsføringen," *Aftenposten*, April 14, 1989.
34. World Commission on Environment and Development, *Our Common Future*; for an analysis of the subsequent discussion of this report, see Guri Bang Søfting and Gro Harlem Brundtland, eds., *The Brundtland Commission's Report—10 Years* (Oslo: Scandinavian University Press, 1998).
35. "Miljøpolitikk i teori og praksis," *Aftenposten*, October 17, 1988.
36. "Borgerlige miljøverntalsmenn: Brundtlandrapport for lite konkret," *Aftenposten*, May 4, 1987.
37. "- Norsk industri er blitt mer miljøbevisst," *Aftenposten*, March 20, 1990.
38. "Fra Ng til G+ i miljøvern," *Aftenposten*, July 12, 1990.
39. "Penger å tjene på miljøvern," *Aftenposten*, March 31, 1989.
40. "Den grønne bølgen er grå," *Aftenposten*, March 18, 1989.
41. Nordic Council of Ministers, *The Swan Label from a Consumer and Environmental Perspective: Evaluation of the Nordic Ecolabeling* (Copenhagen: Nordic Council of Ministers, 2001).
42. Bette K. Fishbein, *Germany, Garbage, and the Green Dot: Challenging the Throwaway Society* (New York: Inform, 1994).

43. European Union, "The European Union Eco-label Homepage," http://ec.europa .eu/environment/ecolabel/.
44. Jan Omdahl, "Grønn økonomidebatt efterlyses," guest editorial, *Aftenposten*, November 27, 1990.
45. Anne Karin Sæther, ed., *Bellona: 20 år i kamp for miljøet* (Oslo: Miljøstiftelsen Bellona, 2006).
46. "Kloster sr. satser på grønn business," *Dagens Næringsliv*, February 23, 1990.
47. "Miljøvernere vil selge kompetanse," *Dagens Næringsliv*, March 17, 1990.
48. "Full rulle i Tomra," *Dagens Næringsliv*, October 7, 1988.
49. "En høyvekst-aksje," *Dagens Næringsliv*, April 14, 1989.
50. "Eventyret fortsetter," *Dagens Næringsliv*, April 14, 1989.
51. "Miljøfondet som slår børsindeksen," *Dagens Næringsliv*, September 3, 1988; "Fond for miljøprofitt," *Dagens Næringsliv*, November 7, 1989.
52. "Dyr og usikker, men kjøp," *Dagens Næringsliv*, March 23, 1990.
53. "Vanskelig å anbefale kjøp," *Dagens Næringsliv*, March 8, 1991.
54. Donella H. Meadows, Dennis L. Meadows, Jørgen Randers, and William W. Behrens III, *The Limits to Growth: A Report for the Club of Rome's Project on the Predicament of Mankind* (London: Pan Books, 1974).
55. "Ønsker å ta noe lite og gjøre det stort," *Aftenposten*, April 4, 1989, morning edition.
56. C. M. Fiol, "A Semiotic Analysis of Corporate Language: Organizational Boundaries and Joint Venturing," *Administrative Science Quarterly* 34 (1989): 277–303. Robert White and Dallas Hanson have argued that "it would be difficult to find an aspect of corporate functioning that has not been studied through annual reports" in "Economic Man and Disciplinary Boundaries: A Case-Study in Corporate Annual Reports," *Accounting, Auditing, and Accountability Journal* 15, no. 4 (2002), 457. Yet I have not found a study of environmental aspects included in annual reports.
57. My approach to companies as storytellers is influenced by David Nye, *Image Worlds: Corporate Identities at General Electric, 1890–1930* (Cambridge: MIT Press, 1985).
58. This was part of a general trend in business. According to Tore Planke, Tomra was one of the first companies in Norway to use Microsoft PowerPoint actively in corporate presentations. Tore Planke, interviewed by author, Oslo, Norway, January 17, 2006, digital recording.
59. Tomra annual report, 1987, 13.
60. Tomra annual report, 1987, 4.
61. Tomra annual report, 1987, 12.
62. Tomra annual report, 1988, 23.
63. Tomra annual report, 1989, 4.
64. Tomra annual report, 1989, 5.
65. Tomra annual report, 1989, 5.
66. Tomra annual report, 1989, 25.
67. Tomra annual report, 1990, 11.
68. "Miljøpris til Asker-bedrift," *NTB Innenriks*, May 8, 1990.
69. *The Economist*, September 1990, quoted in Tomra annual report, 1990, 31.
70. Tomra annual report, 1990, 4.
71. Tomra annual report, 1991, 26–29.
72. Tomra annual report, 1991, 31.
73. Tomra Twenty-fifth anniversary brochure, 11.
74. Tomra annual report, 1991, 10. This wording was slightly modified in the 1992 report: "The company does not pollute the external environment to any greater

extent than is normal for this type of activity," which is a much more realistic statement. It also appears in 1993.

75. Tomra annual report, 1993, 7. An ISO 9001 certification does not in itself guarantee environmentally friendly practices. Rather, the certification attests to the implementation of consistent procedures and monitoring of business processes within the company. International Organization for Standardization, "ISO in brief: International Standards for a Sustainable World," http://www.iso.org/iso/en/aboutiso/isoinbrief/isoinbrief.html. Tomra would later get the more environment- and sustainability-specific ISO 14001 certification.
76. Tomra annual report, 1994, 26.
77. Tomra annual report, 1994, 26. It is interesting that Tomra specifies that its energy comes from hydro-electric power, considering that almost all of the power consumed in Norway is hydro-electric.
78. Tomra annual report, 1995, 26.
79. "Investorene tenker grønt . . . og da kjøper de Tomra," *Finansavisen*, August 15, 1996.
80. Tomra annual report, 1996, 33.
81. Tomra annual report, 1999, 58–64.
82. Tomra annual report, 2000, 31.
83. Tomra annual report, 1999, 64.
84. "Det amerikanske markedet er enormt", *Ingeniørnytt*, no. 14 (1986).
85. Tomra annual report, 1990, 4.
86. "EF avgjør Tomra-vekst," *Dagens Næringsliv*, March 8, 1991.
87. Tomra annual report, 1997, 27.
88. Tomra annual report, 1997, 2.
89. Tomra annual report, 1997, 2.
90. Tomra annual report, 1996, 28.
91. Tore and Petter Planke, interviewed by author, Asker, Norway, June 14, 2005, digital recording.
92. OECD, *Beverage Containers*, 21.
93. "Record: Top condition = top performance and satisfaction," *Return*, no. 1 (2002): 7.
94. Tomra funded significant usability research on the interface before launching the T-600. For instance, Trond Are Øritsland's PhD dissertation from the Department of Product Design, NTNU, used the T-600 interface design as one of his case studies: "Menneskelige aspekter i design: en innføring i ergonomi" (PhD dissertation, Norwegian University of Science and Technology, 1997).
95. Tomra annual report, 1996, 39.
96. Tomra annual report, 1997, 10.
97. "Tomra North America starts coupon test," *Tomra News*, no. 2 (1997): 11.
98. Tomra annual report, 1996, 30.
99. Tomra annual report, 1996, 29.
100. "Tomflasker inn—veldedighet ut," *Aftenposten*, October 16, 1997.
101. "Samler flasker til flyktningene," *Dagens Næringsliv*, November 13, 1996.
102. Tomra annual report, 1996, 28.
103. "Donation projects in Norway," *Tomra News*, no. 2 (1997): 3.
104. "De færreste gir bort flaskepanten," *Aftenposten*, January 11, 1999.
105. "Innsamlingsformen byr kunder imot," *Aftenposten*, January 11, 1999.
106. The cooperation between the three partners in the project—the Salvation Army, the Norwegian Refugee Council, and Norwegian People's Aid—ended on

a sour note when they discovered that the Refugee Council (which initiated the project) had taken seven million kroner in loans on behalf of the two other partners in order to purchase the RVMs. "Armeen fant ikke frelse i returpant," *Dagens Næringsliv*, April 13, 1999.

107. "Gambling, tomflasker og nødhjelp," *Dagens Næringsliv*, April 17, 1999.
108. Tomra annual report, 1996, 30.
109. "Tomra erobrer USA," *Asker og Bærums budstikke*, October 21, 1996.
110. Tomra annual report, 1996, 30.
111. Tomra annual report, 1999, 2.

Chapter 7

1. Odd Børretzen (b. 1926) is a highly respected Norwegian writer, musician, and illustrator known for his warmth and sense of humor. Ravi (Ivar Johansen, b. 1976) is a Norwegian musician, a founding member of Jaga Jazzist, and is now mostly known as a solo artist. Ravi's lyrics are characterized by a creative use of language, spelling, and grammar. The video is available online: http://pant.no/reklamearkiv/ files/movies/resirk_odd_ravi_sommer2006.wmv.
2. The front of the machine does not look like any RVM I have seen, so it is probably a stage-model made just for this video. It is, however, very clearly identifiable as an RVM. By not identifying the machine, Resirk avoided signalling strong connections to any particular RVM manufacturer. Previous Resirk ads featured Tomra machines, for instance the summer 2003 infomercial, also featuring Odd Børretzen.
3. "Pant alt, alltid!" In Norwegian, "pant" is used to mean both the deposit on a bottle (or can) and the act of returning the container to receive your refund.
4. Eric Schatzberg, "Symbolic Culture and Technological Change: The Cultural History of Aluminum as an Industrial Material," *Enterprise and Society* 4, no. 2 (June 2003), 240.
5. "Knuste tomflasker," *VG*, October 3, 1972, 30.
6. "Nei takk til plastflasker," *VG*, October 10, 1972, 30.
7. "'Brusmonopolet' blir angrepet," *Aftenposten*, December 2, 1985, 15.
8. Norges Colonialgrossisters Forbund, *100 år i dagligvarehandelens tjeneste 1908–2008* (Oslo: Norges Colonialgrossisters Forbund, 2008), 25.
9. "Snacks til salgs," *VG*, November 2, 1982, 3.
10. "Mineralvannproduksjonen," *Aftenposten*, May 8, 1984, 14.
11. "'Priskrig' allerede igang: Brus på store plastflasker," *Aftenposten*, May 30, 1984, 14.
12. For comparison, a bottle of the same size today generally costs between eight and twelve kroner. Adjusted for inflation, the 1984 price is equivalent to approximately thirty kroner today. In other words, soda has gotten *much* cheaper the past twenty-five years.
13. "'Brusmonopolet' blir angrepet," *Aftenposten*, December 2, 1985, 15.
14. Resirk, *Pante- og retursystemer på drikkevaresektoren i Norge* (Oslo: Resirk, 1990), 33.
15. Nora had 150,000 shares, making them the sixth-largest shareholder. "Sluttspurt for redning av Tomra," *Aftenposten*, August 7, 1986.
16. Tomra annual report, 1984, 16.
17. Tomra annual report, 1985, 4.
18. For a closer discussion of the Tomra 300, see Chapter 4.

19. "Oslo vil unngå flaskeskår: Fjern avgift på engangsemballasje," *Aftenposten*, July 6, 1985, 12.
20. "'Brusmonopolet' blir angrepet," *Aftenposten*, December 2, 1985, 15.
21. "Bokse-kampen hardner til," *Dagens Næringsliv*, July 8, 1989.
22. "Kjøpmenn vil ha emballasjepant," *Aftenposten*, April 18, 1989, 17.
23. "Kjøpmenn vil ha emballasjepant," *Aftenposten*, April 18, 1989, 17.
24. "Returkaos för plastflaskor" *Göteborgs-Posten*, January 18, 1994, 21.
25. "Tjugo år sedan första pantflaskan," *Gotlands Allehanda*, December 16, 2008, 19.
26. "Pantat agerande," *Aftonbladet*, August 22, 1995, 2.
27. "Returpack betalar inte för burken," *Göteborgs-Posten*, August 12, 1995, 6.
28. "Handels får nya returautomater," *Helsingborgs Dagblad*, August 5, 1995, 21.
29. "Rekord-resultat fra Tomra," *Dagens Næringsliv*, July 14, 1995, 13.
30. "Urdrucken burk är guld vard," *Svenska Dagbladet*, May 2, 1996, 46.
31. "Utländska burkar ersätts inte," *Göteborgs-Posten*, June 9, 1995, 5.
32. "Dubbel mängd utländska ölburkar 1997," *Svenska Dagbladet*, September 16, 1996, 34.
33. "Olaglig läskimport hot mot svenskt pantsystem," *Tidningarnas Telegrambyrå*, September 5, 2001.
34. "Utländska tomburkar på sopberget," *Göteborgs-Posten*, September 26, 1997, 33.
35. "Norge i boks—EF nøler," *Dagens Næringsliv*, November 7, 1989.
36. "Aluminum og plast i europeisk duell," *Dagens Næringsliv*, September 25, 1989.
37. "Aluminum og plast i europeisk duell," *Dagens Næringsliv*, September 25, 1989.
38. Part of my discussion of Resirk is based on Erik Røsrud, *RESIRK historien 1989–1999* (unpublished book manuscript). Thanks to Erik Røsrud for providing me with a copy of his manuscript.
39. Røsrud, *RESIRK historien*.
40. Resirk, *Pante- og retursystemer på drikkevaresektoren i Norge*. The following organizations were behind the proposal: Norges Dagligvarehandels Forbund, Norsk Bryggeri- og Mineralvannindustris Forening, Norges Kooperative Landsforening, Norges Colonialgrossisters Forbund, KØFF-gruppen A/S, Dagligvare Leverandørenes Forening, and Frukt- og Tobakkhandlernes Landsforbund. In addition, the following companies were involved in making the report: Hydro Aluminium A/S, A/S Hansa Nord, Ringes A/S, Joh.Johannson, Tomra Systems A/S, Hagen-gruppen A/S, Gimsøy Kloster A/S, and Egersund Mineralvannfabrikk A/S. As we see, the grocers and the beverage industry dominated the involved actors. Tomra and Hydro Aluminium were the only two participants not belonging to these industries.
41. Resirk, *Pante- og retursystemer på drikkevaresektoren i Norge*, 50
42. "Bokse-kampen hardner til," *Dagens Næringsliv*, July 8, 1989.
43. "Berntsen redder PLM Moss," *Dagens Næringsliv*, January 7, 1993.
44. Coca-Cola Nordic and Northern Eurasia Division to Miljøverndepartementet, December 16, 1994, folder "92/2496 Revidering virkemidler mot emballasje, Mappe III," Ministry of Environment archives.
45. "Hansa tapte boksekampen," *Dagens Næringsliv*, December 13, 1991.
46. NNNs emballasjeutvalg, "Stopp innføringen av retursystemer på engangsemballasje for øl og mineralvann," note, October 1990. Stein Stugu private collection.
47. Biong, *Mulige konsekvenser av å innføre aluminiumsboks som drikkevareemballasje i Norge i forbindelse med et pante- og retursystem: Prosjekt Del I*, 5.
48. Harald Biong, *Mulige konsekvenser av å innføre aluminiumsboks som drikkevareemballasje i Norge i forbindelse med et pante- og retursystem: Prosjekt del II—Holdninger blant norske forbrukere til boks og annen emballasje for øl og mineralvann—kommentarrapport* (Sandvika: Norsk Institutt for Markedsforskning, 1991), 31.

49. "Avgifts-diktert," *Dagens Næringsliv*, November 24, 1993.
50. "Hvis avgiften forsvinner: Nye industrijobber i boks," *Aftenposten*, September 30, 1991, morning edition.
51. Gemini Consulting A/S, *Sysselsettingsmessige virkninger av innføringen av Resirk-systemet for håndtering av engangsemballasje for øl og mineralvann* (Gemini Consulting: Oslo, 1992).
52. "Hvis avgiften forsvinner: Nye industrijobber i boks," *Aftenposten*, September 30, 1991, morning edition.
53. "Bryggeriarbeidere streiket mot å fjerne miljøavgift," *NTB Innenriks*, November 12, 1991.
54. "Engangsemballasje en fordel," *Aftenposten*, November 20, 1991, morning edition.
55. "Naturvernforbundet: - Øl på boks betyr svekket miljøinnsats," *Aftenposten*, May 22, 1993, morning edition.
56. "Furten flaskefabrikk", *Dagens Næringsliv*, June 28, 1993.
57. "Hansa tapte boksekampen," *Dagens Næringsliv*, December 13, 1991.
58. Tomra annual report, 1991, 5.
59. Tomra annual report, 1991, 8.
60. Norway, Ministry of the Environment, *Avfallsminimering og gjenvinning*, NOU 1990:28.
61. Norway, Ministry of the Environment, *Om tiltak for reduserte avfallsmengder, økt gjenvinning og forsvarlig avfallsbehandling*, St. meld. 44, 1991–92.
62. Storting discussion of the 1993 national budget.
63. Norway, Ministry of the Environment, *Avfallsminimering og gjenvinning*, NOU 1990:28, 26.
64. "Tar sysselsettingshensyn i boks-saken," *Dagens Næringsliv*, January 9, 1993.
65. "Tomra-seier i bokse-kamp," *Asker og Bærum Budstikke*, May 13, 1993.
66. "Avgifts-diktert," *Dagens Næringsliv*, November 24, 1993.
67. "Stopp for Resirk," *Dagens Næringsliv*, December 2, 1993.
68. Resirk, press release, February 9, 1994. Quoted in Røsrud, *RESIRK historien*, 13–14.
69. "Berntsen nedstemt om emballasjeavgift," *Dagens Næringsliv*, June 16, 1993.
70. "Grunnavgift skader," *Dagens Næringsliv*, November 26, 1993.
71. "Grunnavgift gir stopp i gjenvinning," *Aftenposten*, December 14, 1993, evening edition.
72. "Flasker i retur," *Dagens Næringsliv*, January 5, 1995.
73. "Staten tapte 'boksekampen'," *Arbeiderbladet*, November 20, 1996.
74. "Berntsen jubler, handelen jubler og Tomra jubler," *Dagligvarehandelen*, December 4, 1996. ESA's task is to ensure that EU competition regulations are followed.
75. "Ny kamp om arbeidsplassene i bryggeribransjen," *Flaskeposten*, September 20, 1995, 1; Stein Stugu to Finanskomiteen og Miljø- og energikomiteen, October 27, 1993, Stein Stugu private collection.
76. The restarted Resirk organization was just one of several voluntary environmental and industry-run agreements initiated by business organizations and individual firms after 1995 to reduce waste. Organizations were founded for the recycling and reclamation of paper, glass, plastic, metals, and beverage containers. These coordinating organizations were established with political and financial support from the Ministry of the Environment.
77. Natur og Ungdom to Miljøverndepartementet/Statens forurensningstilsyn, March 16, 1999; Norsk Nærings- og Nytelsesmiddelarbeiderforbund to Miljøverndepartementet/Statens forurensningstilsyn, March 17, 1999. Stein Stugu private collection.
78. ECON—Senter for økonomisk analyse, *Pant på engangs drikkevare-emballasje* (Oslo: ECON, 1998), 55.

79. Tomra annual report, 1985, 9.
80. Tomra annual report, 1990, 7.
81. "Ønsker pant på bokser," *Aftenposten*, November 1, 1991, evening edition.
82. Røsrud, *RESIRK historien*.
83. Erik Røsrud and Jarle Grytli, interviewed by author, Oslo, Norway, January 16, 2006, digital recording. Tomra's annual report, 1989, contains a model of how a deposit system works, which looks very similar to the Resirk setup.
84. Hughes, *American Genesis*.
85. "Bransjen mangler en god lobbyist," *Dagligvarehandelen*, January 27, 1994.
86. Tomra annual report, 1991, 6.
87. *Tomra News*, April 1993, 3.
88. "Opprykk i Tomra," *Asker og Bærums Budstikke*, April 18, 1996.
89. Tomra annual report, 1997, 2.
90. Jarle Grytli, "Norsk Resirk, finally 'on-line,'" *Return*, no. 1 (1999): 10.
91. "Automatiske utfordrere til Tomra," *Dagens Næringsliv*, September 30, 1997, 15.
92. Tomra annual report, 1997, 24.

Chapter 8

1. Resirk, "Framtidens påskeføre?" print advertisement, *Dagens Næringsliv*, March 31, 2007.
2. Here I use the term *discourse* in a loose sense, as a set of public debates and exchanges surrounding related groups of themes.
3. These discourses obviously live on even though I have chosen to not follow all of the many national and international laws, acts, and regulations that discuss beverage container reuse and recycling in detail.
4. Siegfried Gideon, *Mechanization Takes Command: A Contribution to Anonymous History* (New York: W. W. Norton, 1969), 3.
5. Bruno Latour, *The Pasteurization of French Society, with Irreductions* (Cambridge: Harvard University Press, 1987), 43.
6. Tore Planke, "Packaging and the Environment" (manuscript, 1995). Tore Planke private collection.
7. W. Bernard Carlson, "Artifacts and Frames of Meaning: Thomas Edison, His Managers, and the Cultural Construction of Motion Pictures," in *Shaping Technology/Building Society: Studies in Sociotechnical Change*, ed. Weibe E. Bijker and John Law, 175–198 (Cambridge: MIT Press, 1992).
8. France, National Assembly, Proposition de loi, "Visant à instaurer un système de consigne pour les bouteilles de bière en verre," no. 232, presented by M. Stéphane Demilly, September 27, 2007.
9. Jon Mooallem, "The Unintended Consequences of Hyperhydration," *New York Times*, May 27, 2007.

Bibliography

Archives

Aage Fremstad private collection
National Archival Services of Norway
Norwegian Ministry of the Environment archives
Petter Planke private collection
Tore Planke private collection
Tomra corporate archives

Interviews

INTERVIEWS BY THE AUTHOR

My own interviews were digitally recorded and partially transcribed by myself. Thanks to Kim Preben Larsen and the Centre for Entrepreneurship at the University of Oslo for providing me with a meeting room to conduct interviews in Oslo.

Aage Fremstad, Oslo, Norway, August 1, 2005. Digital recording.
Jarle Grytli, Oslo, Norway, January 16, 2006. Digital recording.
Svein Jacobsen, Billingstad, Norway, January 17, 2006. Digital recording.
Petter and Grete Planke, Vollen, Norway, August 2, 2005. Digital recording.
Tore Planke, Oslo, Norway, October 25, 2005 and January 17, 2006. Digital recording.
Tore and Petter Planke, Asker, Norway, June 14, 2005. Digital recording.
Jørgen Randers, Oslo, Norway, January 12, 2006. Written notes.
Erik Røsrud, Oslo, Norway, January 16, 2006. Digital recording.
Stein Stugu, Gjelleråstoppen, Norway, January 16, 2006. Written notes.
Fredrik Tveitan, August 15, 2006. Telephone interview, written notes.

INTERVIEWS BY CAROL QUINN

The interviews carried out by Carol Quinn took place in 2000 and 2001 as part of an initiative to document Tomra's history. The interviews were conducted in English, recorded on videotape, and transcribed by Norwegian students working for Tomra. In several cases, I have corrected the transcriptions.

Andreas Nordbryhn, August 14, 2000.
Petter and Tore Planke (commercial interview), August 8–9, 2000.
Tore Planke (technical interview), August 9, 2000.
Erik Thorsen and Helge Nerland, April 9, 2001.

Primary and Secondary Sources

Aakvaag, Gunnar C. *Forbrukersamvirket og medlemmene 1970–2004: Mellom sosialdemokratisk modernisering og nyliberal individualisering*. ISF Report 2004:18. Oslo: Institutt for samfunnsforskning, 2004.

Ackerman, Frank. *Why Do We Recycle? Markets, Values, and Public Policy*. Washington, DC: Island Press, 1997.

Allenby, Braden R., and Deanna J. Richards, eds., *The Greening of Industrial Ecosystems*. Washington, DC: National Academy Press, 1994.

Aluminum Company of America. *Aluminum by Alcoa*. Pittsburg, Pa.: Aluminum Company of America, 1969.

Amdam, Rolv Petter, Dag Gjestland, and Andreas Hompland, eds. *Årdal: Verket og bygda 1947–1997*. Oslo: Samlaget, 1997.

Aquino, John T. *Waste Age: Recycling Times' Recycling Handbook*. Washington, DC: Environmental Industry Associations, 1995.

Asdal, Kristin. *Økonomer og miljøavgifter—en historisk analyse*. Rapportserie fra Alternativ Framtid, nr. 6. Oslo: Alternativ Framtid, 1995.

———. "Politikkens teknologier: Produksjoner av regjerlig natur." PhD diss., University of Oslo, 2004.

Baum, Dan. *Citizen Coors: An American Dynasty*. New York: William Morrow, 2002.

Berntsen, Bredo. *Grønne linjer: Natur- og miljøvernets historie i Norge*. Oslo: Grøndal Dreyer, 1994.

Bess, Michael. *The Light-Green Society: Ecology and Technological Modernity in France, 1960–2000*. Chicago: University of Chicago Press, 2003.

Beverage World. *100 Year History of the Beverage Marketplace, 1882–1982 and Future Probe*. New York: Beverage World, 1982.

Bijker, Wiebe. *Bicycles, Bakelites, and Bulbs: Toward a Theory of Sociotechnical Change*. Cambridge: MIT Press, 1995.

Biong, Harald. *Mulige konsekvenser av å innføre aluminiumsboks som drikkevareemballasje i Norge i forbindelse med et pante- og retursystem: Prosjekt Del I—en studie av aluminiumsboksens utvikling i Sverige*. Sandvika: Norsk Institutt for Markedsforskning, 1990.

———. *Mulige konsekvenser av å innføre aluminiumsboks som drikkevareemballasje i Norge i forbindelse med et pante- og retursystem: Prosjekt del II—Holdninger blant norske forbrukere til boks og annen emballasje for øl og mineralvann—kommentarrapport*. Sandvika: Norsk Institutt for Markedsforskning, 1991.

Blomkvist, Pär, and Arne Kaijser. *Den konstruerade världen: Tekniska system i historisk perspektiv*. Stockholm: Brutus Östlings Bokförlag Symposion, 1998.

Blumberg, Louis, and Robert Gottlieb. *War on Waste: Can America Win Its Battle with Garbage?* Washington, DC: Island Press, 1989.

Bryggeri- og mineralvannfabrikkarbeidernes forening. *"Vårt bidrag": Bryggeri- og mineralvannfabrikkarbeidernes forening gjennom 100 år: 1884–1984*. Oslo: Bryggeri- og mineralvannfabrikkarbeidernes forening, 1984.

Carlson, W. Bernard. "Artifacts and Frames of Meaning: Thomas Edison, His Managers, and the Cultural Construction of Motion Pictures." In *Shaping Technology/Building Society: Studies in Sociotechnical Change*, edited by Weibe E. Bijker and John Law, 175–198. Cambridge: MIT Press, 1992.

Chandler, Alfred D., Jr. *Strategy and Structure: Chapters in the History of the Industrial Enterprise*. Cambridge: MIT Press, 1962.

City of New York Department of Sanitation. *Public Comments on a Comprehensive Solid Waste Management Plan for New York City and Draft Generic Environmental Impact Statement Submitted in June 1992*. New York: City of New York Department of Sanitation, 1992.

Dammann, Erik. *Fremtiden i våre hender: Om hva vi alle kan gjøre for å styre utviklingen mot en bedre verden*. Oslo: Gyldendal, 1972.

E. C. Dahls bryggeri. *Aksjeselskapet E. C. Dahls bryggeri 1856–1956: 100 år*. Trondheim: E. C. Dahl, 1956.

ECON—Senter for økonomisk analyse. *Pant på engangs drikkevare-emballasje*. Oslo: ECON, 1998.

Ellis, William S. *Glass: From the First Mirror to Fiber Optics, The Story of the Substance That Changed the World*. New York: Avon Books, 1998.

Erichsen, Eivind. "Må vi velge mellom økonomisk vekst og miljøvern?" In *Økonomi og politikk: 15 artikler*, edited by Petter Jakob Bjerve and Ole Myrvoll, 51–66. Oslo: Aschehoug, 1971.

Fallan, Kjetil. "How an Excavator Got Aesthetic Pretensions—Negotiating Design in 1960s' Norway." *Journal of Design History* 20, no. 1 (2007): 43–59.

Fiol, C. M. "A Semiotic Analysis of Corporate Language: Organizational Boundaries and Joint Venturing." *Administrative Science Quarterly* 34 (1989): 277–303.

Fishbein, Bette K. *Germany, Garbage, and the Green Dot: Challenging the Throwaway Society*. New York: Inform, 1994.

France, National Assembly. Proposition de loi, "Visant à instaurer un système de consigne pour les bouteilles de bière en verre," no. 232, presented by M. Stéphane Demilly, September 27, 2007.

Friedel, Robert. "A New Metal! Aluminum in its 19th-Century Context." In *Aluminum By Design*, edited by Sarah Nichols, 59–83. Pittsburgh: Carnegie Museum of Art, 2000.

Gandy, Matthew. *Recycling and the Politics of Urban Waste*. New York: St. Martin's Press, 1997.

Gemini Consulting A/S. *Sysselsettingsmessige virkninger av innføringen av Resirk-systemet for håndtering av engangsemballasje for øl og mineralvann*. Oslo: Gemini Consulting, 1992.

Gideon, Siegfried. *Mechanization Takes Command: A Contribution to Anonymous History*. New York: W. W. Norton, 1969.

Golding, Andreas. *Reuse of Primary Packaging, Final Report*. Study B4–3040/98/0001080/MAR/E3. Brussels: European Commission Environment Directorate-General, 1998.

Gould, Kenneth A., Allan Schnaiberg, and Adam S. Weinberg, *Local Environmental Struggles: Citizen Activism in the Treadmill of Production*. Cambridge: Cambridge University Press, 1996.

Grossman, Elizabeth. *High Tech Trash: Digital Devices, Hidden Toxics, and Human Health*. Washington, DC: Island Press, 2006.

Guber, Deborah Lynn. *The Grassroots of a Green Revolution: Polling America on the Environment*. Cambridge: MIT Press, 2003.

Hajer, Maarten A. *The Politics of Environmental Discourse: Ecological Modernization and the Policy Press*. Oxford: Clarendon Press, 1995.

Hart, Susannah, and John Murphy, eds. *Brands: The New Wealth Creators*. Basingstoke, U.K.: Palgrave, 1998.

Heritage of Splendor. Film. Produced for Keep America Beautiful by Richfield Oil Corporation. Hollywood, Calif.: Higgins (Alfred) Productions, 1963.

Hickman, H. Lanier, Jr. *American Alchemy: The History of Solid Waste Management in the United States*. Santa Barbara, Calif.: Forester Press, 2003.

Hine, Thomas. *The Total Package: The Secret History and Hidden Meanings of Boxes, Bottles, Cans, and Other Persuasive Containers*. Boston: Little, Brown, 1997.

Hughes, Thomas P. *American Genesis: A Century of Invention and Technological Enthusiasm, 1870–1970*. New York: Viking, 1989.

———. *The Human-Built World: How to Think about Technology and Culture*. Chicago: University of Chicago Press, 2004.

———. *Networks of Power: Electrification in Western Society, 1880–1930*. Baltimore: Johns Hopkins University Press, 1983.

Jacobsen, Stian. *Norges beste gründere—og floppene*. Oslo: Hegnar Media, 2002.

Jamison, Andrew. *The Making of Green Knowledge: Environmental Politics and Cultural Transformation*. Cambridge: Cambridge University Press, 2001.

Jørgensen, Finn Arne. "Keep America Beautiful." In *Encyclopedia of American Environmental History*, edited by Kathleen Brosnan. New York: Facts on File, 2010.

Kelly, Katie. *Garbage: The History and Future of Garbage in America*. New York: Saturday Review Press, 1973.

Killengreen, Chr. P. *Den Norske Bryggeriforening: Jubileumsskrift i anledning foreningens 25 aars bestaaen*. Oslo: Centraltrykkeriet, 1926.

Kiuchi, Tachi, and William K. Shireman. *What We Learned in the Rainforest: Business Lessons from Nature*. San Francisco: Berrett-Koehler, 2002.

Koppell, G. Oliver. *Inconspicuous Consumption: A Look at New York's Bottle Bill and the Experience of Six States, which have Enacted Returnable Container Legislation*. Albany, N.Y.: New York State Assembly, 1981.

Kranzberg, Melvin. "Technology and History: 'Kranzberg's Laws.'" *Technology and Culture* 27 (1986): 544–560.

Latour, Bruno. *The Pasteurization of French Society, with Irreductions*. Cambridge: Harvard University Press, 1987.

Lidgren, Karl. *Dryckesförpackningar och miljöpolitik—en studie av styrmedel*. Lund: Lund Economic Studies, 1978.

Lowry, Edwin F., Thomas W. Fenner, and Rosemary M. Lowry. *Disposing of Nonreturnables: A Guide to Minimum Deposit Legislation*. Stanford, Calif.: Stanford Environmental Law Society, 1975.

Lynskey, Michael J. Introduction to *Entrepreneurship and Organization: The Role of the Entrepreneur in Organizational Innovation*, edited by Michael J. Lynskey and Seiichiro Yonekura, 1–57. Oxford: Oxford University Press, 2002.

MacFarlane, Alan, and Gerry Martin. *Glass: A World History*. Chicago: University of Chicago Press, 2002.

McEwen, L. B. *Waste Reduction and Resource Recovery Activities: A Nationwide Survey*. Washington, DC: U.S. Environmental Protection Agency, 1977.

Meadows, Donella H., Dennis L. Meadows, Jørgen Randers, and William W. Behrens III. *The Limits to Growth: A Report for the Club of Rome's Project on the Predicament of Mankind*. London: Pan Books, 1974.

Melosi, Martin V. *Garbage in the Cities: Refuse, Reform, and the Environment: 1880–1980*. College Station: Texas A&M University Press, 1981.

———. *The Sanitary City: Urban Infrastructure in America from Colonial Times to the Present*. Baltimore: Johns Hopkins University Press, 2000.

Miller, Benjamin. *Fat of the Land: Garbage in New York the Last 200 Years*. New York: Four Walls Eight Windows, 2000.

Myrvang, Christine, Sissel Myklebust, and Brita Brenna. *Temmet eller uhemmet: Historiske perspektiver på konsum, kultur og dannelse*. Oslo: Pax Forlag, 2004.

Nordic Council of Ministers. *The Swan Label from a Consumer and Environmental Perspective: Evaluation of the Nordic Ecolabeling*. Copenhagen: Nordic Council of Ministers, 2001.

Norges Colonialgrossisters Forbund. *100 år i dagligvarehandelens tjeneste 1908–2008*. Oslo: Norges Colonialgrossisters Forbund, 2008.

Norsk Bryggeri- og mineralvannsforening. *Miljøhåndbok for bryggeribransjen 1995*. Oslo: Norsk Bryggeri- og mineralvannsforening, 1995.

Norway. *Lov om adgang til å forby bruken av visse slag engangsemballasje ved markedsføring av forbruksvarer*. Ot. prp. nr. 77, 1969–1970.
———. *Lov om naturvern*. Ot. prp. nr. 65, 1968–69, June 19, 1970.
———. *Lov om produktkontroll*. Ot. prp. nr. 51, 1974–75, April 18, 1975.
———. *Om midlertidig lov om panteordninger for emballasje til øl, mineralvann og andre leskedrikker*. Ot. prp. nr. 61, 1973–1974, May 27, 1974.
———. *Statsbudsjettet*. St. prp. no. 1, 1973–74.
Norway, Ministry of the Environment. *Avfallsminimering og gjenvinning*. NOU 1990:28.
———. *Om tiltak for reduserte avfallsmengder, økt gjenvinning og forsvarlig avfallsbehandling*. St. meld. 44, 1991–1992.
———. *Resirkulering og avfallsbehandling II*. NOU 1975:52.
Norway, Ministry of Finance. "Forskrifter om panteordninger for emballasje til øl, mineralvann og andre leskedrikker: Forskrifter om avgift på øl og kullsyreholdige, alkoholfrie drikkevarer i engangsemballasje,"14 June 1974.
Norway, Odelsting discussion. "Diskusjon av Ot. prp. nr. 77, 1969–70." O. tid., 485–487, 1970.
Norway, Storting discussion. "Innstilling fra finanskomiteen om midlertidig lov om panteordninger for emballasje til øl, mineralvann og andre leskedrikker." Innst. O. Nr. 57. S. tid. 651, 1974.
Nye, David. *Image Worlds: Corporate Identities at General Electric, 1890–1930*. Cambridge: MIT Press, 1985.
Ødegaard, Arne, Jan Grønli, and Ole-Jørgen Lier. *PLM Moss glassverk AS 1898–1998*. Moss: Glassverket, 1998.
OECD. *Beverage Containers: Re-use or Recycle*. Paris: Organization for Economic Co-operation and Development, 1978.
Øritsland, Trond Are. "Menneskelige aspekter i design: en innføring i ergonomi." PhD dissertation, Norwegian University of Science and Technology, 1997.
Østby, Per. "Flukten fra Detroit: Bilens integrasjon i det norske samfunnet." PhD dissertation, University of Trondheim, 1995.
Overbye, Signy Ryther. "Etableringen av en norsk skipsautomatiseringsindustri." In *Elektronikkentreprenørene: Studier av norsk elektronikkforskning og–industri etter 1945*, edited by Olav Wicken, 152–177. Oslo: Ad Notam Gyldendal, 1994.
Øystå, Øystein. *Brygg, brus og bruduljer: Bryggeri- og mineralvannbransjen i Norge 100 år*. Oslo: Bryggeri- og mineralvannforeningen, 2001.
Pendergrast, Mark. *For God, Country, and Coca-Cola: The Unauthorized History of the Great American Soft Drink and the Company That Makes It*. New York: Basic Books, 1993.
Petersen, E., and C. J. Arnholm. *Frydenlund Bryggeri: 100 år 1859–1959*. Oslo: Frydenlund Bryggeri, 1959.
Planke, Sverre Martens. *Det var en gang: Erindringer fra et helt århundre*. Tvedestrand: Toring AS, 2001
Rathje, William, and Cullen Murphy. *Rubbish! The Archeology of Garbage*. Tucson: University of Arizona Press, 2001.
Resirk. *Pante- og retursystemer på drikkevaresektoren i Norge*. Oslo: Resirk, 1990.
Rigello Pak AB. *Basfakta i debatten miljövård—dryckesförpackningar*. Lund: Rigello Pak, 1969.
———. *Dryckesförpackningarna och energin = Beverage containers and energy = Getränkeverpackungen und die Energie = Emballages de boisson et consommation d'énergie*. Lund: Rigello Pak, 1974.
———. *Plastics from an Environmental Standpoint: Kunststoffe und unsere Umwelt = Matières plastiques et l'environnement*. Lund: Rigello Pak 1970.
Rogers, Heather. *Gone Tomorrow: The Hidden Life of Garbage*. New York: New Press, 2006.

Rothman, Hal K. *The Greening of a Nation? Environmentalism in the United States since 1945*. Fort Worth, Tex.: Harcourt Brace College Publishers, 1998.

Rønningsbakk, Bernt. "Nyskaping og dialog: Med 12 norske eksempler." Doctoral thesis, Norges Tekniske Høyskole, 1995.

Røsrud, Erik. *Dagligvareforbundets rolle og virke mot år 2000*. Oslo: Dagligvareforbundet, 2003.

———. "RESIRK historien 1989–1999." Manuscript.

Royte, Elizabeth. *Garbage Land: On the Secret Trail of Trash*. Boston: Little, Brown, 2005.

Sæther, Anne Karin, ed. *Bellona: 20 år i kamp for miljøet*. Oslo: Miljøstiftelsen Bellona, 2006.

Segrave, Kerry. *Vending Machines: An American Social History*. Jefferson, N.C.: McFarland, 2002.

Schatzberg, Eric. "Symbolic Culture and Technological Change: The Cultural History of Aluminum as an Industrial Material," *Enterprise and Society* 4, no. 2 (June 2003): 226–271.

Schreiber, G. R. *A Concise History of Vending in the U.S.A.* Chicago: Vend, 1961.

Schulerud, Mentz. *Ringnes bryggeri gjennom 100 år*. Oslo: Rignes, 1976.

Schumpeter, Joseph A. "Economic Theory and Entrepreneurial History." In *Essays of J. A. Schumpeter*, edited by R. V. Clemence, 248–266. Cambridge, Mass.: Addison-Wesley Press, 1951.

———. *The Theory of Economic Development: An Inquiry into Profits, Capital, Credit, Interest, and the Business Cycle*. Cambridge: Harvard University Press, 1934.

Seinfeld. "The Bottle Deposit." Parts 1 and 2, season 7, episodes 21 and 22. NBC. Aired May 2, 1996.

Shireman, William K. *The CalPIRG-ELS Study Group Report on Cans and Bottle Bills*. Stanford, Calif.: Stanford Environmental Law Society, 1981.

Shiva, Vandana. *Water Wars: Privatization, Pollution, and Profit*. London: Pluto Press, 2002.

Skjåk, Knut Kalgraff, and Bjug Bøyum. *Undersøking om verdier, natur og miljø 1993*. Oslo: Norsk Samfunnsvitenskapelig Datatjeneste, 1993.

Smith, George David. *From Monopoly to Competition: The Transformations of Alcoa, 1888–1986*. Cambridge: Cambridge University Press, 1986.

Søfting, Guri Bang, and Gro Harlem Brundtland, eds. *The Brundtland Commission's Report—10 Years*. Oslo: Scandinavian University Press, 1998.

Strand, Arvid. *Dagligvarekjøpmennene og samfunnet 1958–1988*. Oslo: Norges dagligvarehandels forbund, 1988.

Strasser, Susan. *Waste and Want: A Social History of Trash*. New York: Henry Holt, 1999.

Stølsbotn, Kristin. *Undersøkelse om verdier, natur og miljø 2000*. Oslo: Norsk Samfunnsvitenskapelig Datatjeneste, 2002.

Sverrison, Árni. "Translation Networks, Knowledge Brokers, and Novelty Construction: Pragmatic Environmentalism in Sweden," *Acta Sociologica* (2001): 313–327.

Sweden, Ministry of Agriculture. *Dryckesförpackningar och miljö*, SOU 1976:35.

———. *Utredning om frågor om återvinning av dryckesförpackninger*, DS Jo 1980:05.

Sweden, Ministry of the Environment. *Översyn av skatten på dryckesförpackningar*, SOU 1990:85.

Sweden, Ministry of Heath and Social Affairs. *Ett renare samhälle*, SOU 1969:18.

Sweden, Riksdagen. "Avgift på vissa dryckesförpackningar." Prot. 1973:27, February 21, 1973.

———. *Lag om återvinning av dryckesförpackningar av aluminium*, 1982:349.

———. *Om återvinning av dryckesförpackningar av aluminium*. Proposition 1981/82:131, February 24, 1982.

———. "Om PLMs tillverkning av aluminiumburkar i Fosie." Prot. 1980/81:94, March 12, 1981.

———. "Om tillverkningen av öl- och läskburkar av aluminium." Prot. 1979/80:42, December 3, 1979.

———. *Regeringens proposition om återvinning och omhändertagande av avfall*. Proposition 1975:32 February 27, 1975.

———. "Svar på fråga 1984/85:384 om insamlingen av aluminiumburkar." Prot. 1984/85:87, February 21, 1985

Sweden, Tax Committee. "Om återvinning av dryckesförpackningar av aluminium (prop. 1981/82:131)." SkU 1981/82:61, April 27, 1982.

Szasz, Andrew. *EcoPopulism: Toxic Waste and the Movement for Environmental Justice*. Minneapolis: University of Minnesota Press, 1994.

Tomra Can-Can. Promotional video. Asker, Norway: Tomra Systems, 1984.

U.S. Resource Conservation Committee. *Briefing Book on Beverage Container Deposit Policy*. Background paper no. 1. Washington, DC, 1977.

Ulleberg, Karl H. *Automathandelen i Norge og automater som omsetningsform*. Bekkestua: Norges Handelshøyskole, 1968.

United States Geological Service. *Aluminum Statistics: 1900 to 2002*. Denver, Colo.: USGS, August 2004.

Vesilind, Aarne, and Thomas Distefano. *Controlling Environmental Pollution: An Introduction to the Technologies, History, and Ethics*. Lancaster, Pa.: DEStech, 2005.

Walth, Brent. *Fire at Eden's Gate: Tom McCall and the Oregon Story*. Portland: Oregon Historical Society Press, 1994.

Whitaker, Jennifer Seymour. *Salvaging the Land of Plenty: Garbage and the American Dream*. New York: William Morrow, 1994.

White, Robert, and Dallas Hanson. "Economic Man and Disciplinary Boundaries: A Case-Study in Corporate Annual Reports." *Accounting, Auditing and Accountability Journal* 15, no. 4 (2002): 450–477.

Winner, Langdon. *The Whale and the Reactor: A Search for Limits in an Age of High Technology*. Chicago: University of Chicago Press, 1986.

World Commission on Environment and Development. *Our Common Future*. New York: Oxford University Press, 1987.

Yearley, Steven. *The Green Case: A Sociology of Environmental Issues, Arguments, and Politics*. London: HarperCollinsAcademic, 1991.

Zimring, Carl A. *Cash for Your Trash: Scrap Recycling in America*. New Brunswick: Rutgers University Press, 2005.

Periodicals

Aftenposten
Aftonbladet
Arbeider-Avisa
Arbeiderbladet
Arbeidsgiveren
Asker og Bærum Budstikke
Dagbladet
Dagens Næringsliv
Dagens Nyheter
Dagligvarehandelen

Economist
Eksport aktuelt
Elektro
Elektronikk
Finansavisen
FK
Flaskeposten
Fritt Kjøpmansskap
Gjengangeren
Gotlands Allehanda
Göteborgs-Posten
Handelsnytt
Helsingborgs Dagblad
Ingeniørnytt
International Herald Tribune
Kapital
Kapital Data
Kjøpmannsnytt
Ledelse
Morgenposten
Nå
New York Times
Newsweek
Norges Handels og Sjøfartstidende
Norges Industri
Norges Kjøbmannsblad
NTB Innenriks
Økonomisk Rapport
Ottens-kuriren
Svenska Dagbladet
Teknisk Ukeblad
Tidningarnas Telegrambyrå
Time
Tromsø
VG

Tomra Corporate Publications

Advertising materials
Quarterly reports and annual reports (1972–2006)
Return (2000–2006)
Tomra News (1987–1999)

Index

About the Author

FINN ARNE JØRGENSEN is an associate senior lecturer in the history of technology and the environment in the Department of Historical, Philosophical, and Religious Studies at Umeå University, Sweden.